Stephen Moss

UNDERSTANDING BIRD BEHAVIOUR

BLOOMSBURY
LONDON • NEW DELHI • NEW YORK • SYDNEY

To Suzanne, who opened my eyes to the wonders of bird behaviour

Bloomsbury Natural History
An imprint of Bloomsbury Publishing Plc

50 Bedford Square	1385 Broadway
London	New York
WC1B 3DP	NY 10018
UK	USA

www.bloomsbury.com

First published 2015

British Library Cataloguing-in-Publication Data
A catalogue record for this book is available from the British Library.

Library of Congress Cataloguing-in-Publication data has been applied for.

ISBN: PB: 978-1-4729-1206-0
ePDF: 978-1-4729-2585-5
ePub: 978-1-4729-2586-2

2 4 6 8 10 9 7 5 3 1

Designed and typeset in UK by Susan McIntyre
Printed in China

UNDERSTANDING
BIRD BEHAVIOUR

The Wildlife Trusts

The Wildlife Trusts are the UK's largest people-powered organisation caring for all nature – rivers, bogs, meadows, forests, seas and much more. We are 47 Wildlife Trusts covering the whole of the UK with a shared mission to restore nature everywhere we can and to inspire people to value and take action for nature for future generations.

Together we care for thousands of wild places that are great for both people and wildlife. These include more than 760 woodlands, 500 grasslands and even 11 gardens. You're away from your nearest Wildlife Trust nature reserve and most people have one within a few miles of their home.

Our goal is nature's recovery – on land and at sea. To achieve this we rely on the vital support of our 800,000 members, 40,000 volunteers, donors, corporate supporters and funders. To find the Wildlife Trust that means most to you and lend your support, visit wildlifetrusts.org/your-local-trust

Importantly, we encourage people to experience wildlife for themselves. We believe that a deeper appreciation for nature conservation can start with a book such as this one by Stephen Moss. We need more people to understand and value the birds and other wildlife that are to be found in our countryside.

Understanding Bird Behaviour introduces the reader to the habitats and characteristics of birds in life – how and why birds feed, preen, and react with others of their kind. The author compares such behaviour as courtship, fledging, flight and migration among many different species, and investigates the instincts and circumstances that trigger these behaviours.

Few realise just how endangered much of our British wildlife is. In recent years, once-common bird species such as the sparrow and Starling have declined, mainly due to the demands of that modern human living has placed on habitats.

The Wildlife Trusts believe, however, that it is not too late. Much can still be done to reverse the losses of the past, and we all have a part to play in making this happen. One way is to contact your local Wildlife Trust for information on wildlife activities and volunteering opportunities, and on local wild places. Help us to protect wildlife for the future and become a member today! Visit www.wildlifetrusts.org for further information. The Wildlife Trusts is a registered charity (number 207238).

We hope that, with the help of this book, you have fun learning more about birds and their behaviour!

CONTENTS

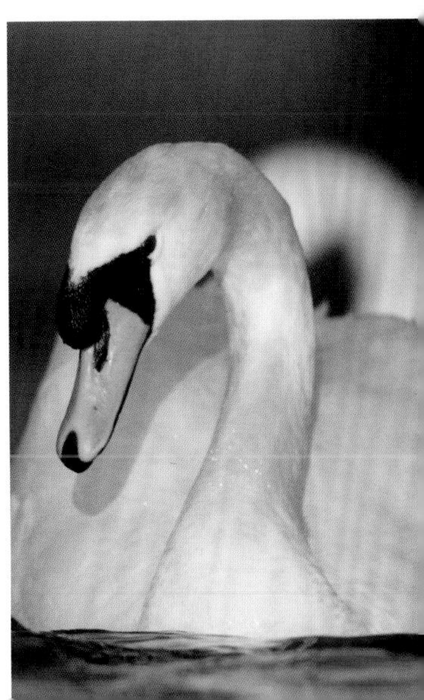

PART TWO: **FAMILIES AND SPECIES** 84

Introduction

Studying bird behaviour is one of the most fascinating and potentially rewarding aspects of watching birds. But where do you start? At first, understanding what birds are doing, and more importantly why, can be confusing – especially if you are a newcomer to birding. Is the aspect of behaviour you are witnessing a normal part of daily life, or something unusual? Will your presence disturb the bird and force it to behave out of character? And how do you interpret some new or different aspect of behaviour you have not witnessed before?

Hopefully, this book will provide some of the answers to these and many other questions. Its purpose is threefold:

i) To provide an introduction to the various different forms and aspects of bird behaviour, categorised by subject

▼ *Snow Buntings often spend the winter on shingle beaches*

ii) To indicate specific types of behaviour characteristic of certain species or family groups
iii) To be a work of reference – use the index to look up either a particular species, or a specific aspect of bird behaviour

The book is divided into two parts, each of which can be read independently, or you can cross-refer between them.

Part 1 covers the various types of bird behaviour, such as flight, courtship, predation and migration, loosely grouped under the chapter headings of **Movement**, **Feeding**, **Breeding**, **Migration and Navigation**, **Distribution and Range**, and **Life and Death**. This provides a quick and easy reference to specific behaviours.

Part 2 is arranged by families, or groups of similar birds, such as seabirds. This includes the 200 or so species that you are most likely to encounter in Britain, some common, others scarce, with details of behaviour common to a particular species or group. This enables you to look up particular species and get some insight into their behaviour, though for reasons of space this cannot possibly be comprehensive. If you are interested in following up the behaviour of a particular species or family, details of suitable works can be found in the Further Reading section at the back of the book.

▲ *Blackbirds sing on spring evenings from the roofs of houses*

▲ *A pair of Fulmars during their elaborate courtship display*

▶ *Like many hole-nesting birds, Blue Tits readily take to using nestboxes*

For much of the second half of the 20th century, thanks to major advances in field guides and optics, birding focused mainly on two related aspects: identification and rarities. While these are both fascinating and important, for a long period they came to dominate birders' minds at the expense of another vital aspect of understanding and enjoying birds, that of studying and interpreting their various behaviours.

With the recent gradual decline of rarity-ticking ('twitching'), and a more enlightened attitude towards the pastime of birding as a way of getting back in touch with nature, the study of bird behaviour is due a renaissance. This book is a small contribution towards encouraging all birders to take a closer interest in what to me personally is by far the most fascinating aspect of watching birds: their behaviour.

Code of Conduct

When watching birds, the welfare of the bird must always come first. Deliberately flushing a bird in order to identify it, getting too close in order to get a good view or a photograph, or disturbing a nesting bird are all unacceptable. Indeed causing disturbance to a Schedule 1 breeding bird at or near the nest is not just wrong but illegal. (For details of birds on Schedule 1 visit the RSPB website.)

▼ *Redwings often visit gardens during hard winter weather, to feed on windfall apples*

PART ONE:

TYPES OF BIRD BEHAVIOUR

The first half of this book deals with the different types of bird behaviour. For the sake of convenience, these are divided into six chapters, each of which deals with a range of related aspects of behaviour:

1 MOVEMENT
- Feathers and flight
- Swimming and diving
- Walking and running
- Flocking
- Roosting and sleeping

2 FEEDING
- Food types and feeding methods
- Predators
- Specialist feeders
- Drinking

3 BREEDING
- Timing
- Territory and song
- Courtship, display and mating
- Nest-building
- Egg laying and incubation
- Parental care and fledging
- Hybridisation
- Unusual breeding behaviour
- Polygamy

4 MIGRATION AND NAVIGATION
- Why do birds migrate?
- How do birds navigate?
- Migration strategies
- Unusual migration

5 DISTRIBUTION AND RANGE
- Habitats and their influence on behaviour
- Range

6 LIFE AND DEATH
- Moult
- Bathing, preening and feather care
- Sight, hearing and smell
- Excretion
- Temperature regulation
- Birds and weather
- Disease and death

To find information about particular species you may also want to refer to the second half of the book, which deals with behaviour on a family-by-family basis.

Chapter 1

Movement

Feathers and flight

One of the characteristics of birds that separates them from most other animals (apart, of course, from bats and many insects) is their ability to fly. They are able to do so because of their unique body structure: a light, hollow skeleton supporting feathers, enabling them to get and stay airborne. That said, different groups of birds have evolved many different ways to fly, including soaring, gliding and flapping with the help of the wind, and using air currents such as thermals.

Birds are able to fly because over many millions of years their bodies have undergone particular adaptations: most importantly the evolution of feathers. These light, versatile structures, probably evolved from reptilian scales, were present in some dinosaurs and today are unique to birds. The flight feathers in the bird's wings and tail are stiff and long, enabling birds to gain and maintain lift and manoeuvre themselves through the complexities of air currents once aloft. In addition, a bird's skeleton is also highly adapted to flight, with strong but hollow bones carrying the minimum of extra weight, meaning that birds are by far the lightest animals for their body size. A **Mute Swan**, for example – one of the world's heaviest flying birds – weighs a mere 10–15kg compared with well over 100kg for a similar-sized mammal.

▼ *A Great Shearwater lives up to its name, gliding low over the waves*

The classic flight mechanism is flapping: moving the wings up and down to gain lift. It is generally used for short, direct flights – for example, a songbird moving from tree to tree – as it consumes a lot of energy. Once aloft, or when travelling for any distance, most birds prefer to use less energy-expensive methods of moving through the air, including gliding and soaring. Seabirds such as **albatrosses** and **shearwaters** are the world's greatest gliders, taking advantage of updraughts from the ocean surface to maintain their position just above the waves, where they can move forward using minimal energy, and hardly flapping their wings for many hours on end. Raptors such as **hawks**, **buzzards** and **eagles** also use gliding flight, during which they put their wings in a position which reduces surface area and allows rapid forward movement while maintaining lift.

▼ *Raptors such as these Common Buzzards use various flying techniques, including flapping, soaring and hovering*

▲ *Sparrowhawks have a characteristic flight style, alternating a series of flaps with a short glide*

Raptors also spend much of the time soaring, a flight style particularly common amongst large, heavy birds, which otherwise would struggle to stay airborne for any length of time. When soaring, a bird like a **Common Buzzard** will spread its wings as wide as possible, maximising their surface area, then take advantage of thermal currents of rising warm air to gain height. Once aloft, it can circle around for some time, again using the minimum of energy. Soaring is generally used to maintain altitude rather than move any distance.

To see the difference, watch a **Sparrowhawk** as it soars overhead on broad, outstretched wings; then see how it changes its wing angle, narrows the wings, and glides rapidly across the sky, appearing quite different in shape from before.

Scientists have compared flight with other mechanisms of locomotion such as walking or swimming, and revealed that it is extraordinarily efficient in comparison. For example, Usain Bolt – the fastest sprinter on the planet – covers the ground at about five body-lengths per second, while the world's fastest land animal, the Cheetah, can manage 18 body-lengths per second. But a flying bird can reach up to 70 or even 80 body-lengths per second, comparable with a jet aircraft. This not only allows birds to get to their destination quickly, it also enables them to cover vast distances, especially on migration, when even a small bird such as a **Swallow** needs to travel thousands of kilometres.

Swimming and diving

Of course, not all birds spend the majority of time in the air. Many waterbirds live most of their lives in an aquatic habitat, and have adapted their physiology and behaviour accordingly. At its simplest, this involves adapting flight techniques when swimming underwater – watch film of **auks** or **penguins** underwater and you will see what I mean. Their wings, which look pretty useless above the surface, are transformed into efficient propellers, enabling them to cover distance and manoeuvre themselves in a very different, and much denser, medium than the air.

Other birds stay mainly on the surface, or divide their time between land and water. Wildfowl such as **ducks**, **geese** and **swans** have adapted to a life spent on water by evolving a number of features, such as fully webbed feet, and oil glands with which they waterproof their feathers on a regular basis to keep them sleek and in good condition. Other waterbirds have different adaptations to their aquatic lifestyle: **grebes** and **Coots** have only partly webbed feet, while **Cormorants** do not have waterproof feathers, so have to stand for long periods drying their wings in their characteristic outstretched pose. However, the lack of waterproofing is an advantage when they dive in shallow water, as they are less buoyant than ducks, so are able to fish in shallow waters without being pushed upwards to the surface.

Waterbirds from unrelated groups often show very similar features and superficial appearance, so **Coots** and **Moorhens** look more like ducks than like other members of their family, the rails and crakes. This is due to a process called convergent evolution, in which external factors dictate the morphology of a bird or other organism, leading it to superficially resemble unrelated species that share the same habitat and lifestyle.

▼ *Gannets feed by plunge-diving straight down into the water at high speed*

Diving is another behaviour shown by different, unrelated groups of birds, including **divers**, **grebes**, many kinds of **ducks**, and seabirds such as **auks**. Again, these species share similar features: a long, streamlined body, webbed or partially webbed feet for underwater propulsion, and legs situated well back on their body, meaning that for most of these birds, getting about on land proves very difficult indeed.

▲ *Auks like this Razorbill appear clumsy in the air, but their short wings are excellent for swimming underwater*

Walking and running

For many landbirds, the best way to travel short distances without using too much precious energy is also the simplest: to walk, hop or run. This is especially true of ground-feeding birds, such as **finches**, **sparrows** and **thrushes**. Different groups use different methods: so tree-dwelling birds such as finches tend to hop when on the ground, while **larks** and **pipits**, which spend much more of their time on the ground itself, will walk or run.

Other groups, such as **gamebirds**, **rails** and **crakes**, spend the vast majority of time on the ground, and are well adapted to running and walking, often hiding in dense cover and only using flight as a last resort. One North American gamebird, the **Mountain Quail**, even migrates on foot, travelling several thousand metres down the mountain to the valleys in autumn, and back again the following spring.

▼ *Red-legged Partridges spend much of their lives on the ground, walking and running*

Flocking

Birds are, by and large, sociable creatures, and many different species gather in flocks, either occasionally or on a regular basis. Flocks occur for various reasons: for example, to maximise chances of finding food, to keep warm, or to avoid predators. It is important to understand that in all these cases the urge to flock must come from each individual, and the advantages of joining a flock must outweigh the disadvantages (such as increased competition) for that individual, rather than the group as a whole.

Flocking also occurs as a by-product of the concentration of food supplies. For example, a rubbish tip will attract the local **gulls**, which will fiercely compete for the available food.

Flocking is much more widespread outside the breeding season, for the simple reason that during the nesting period most birds are in pair bonds with one or more members of the opposite sex, and duties such as incubation and feeding the young preclude joining others in flocks. Once breeding is over, songbirds often gather in small family parties or loose flocks of a dozen or more, but while food is plentiful and there is plenty of cover to avoid predators, there is no pressing need to form larger groups.

Once winter arrives, however, factors such as the reduced amount of daylight in which to find food motivate individuals to seek out others. So from late autumn you often see flocks of **tits** travelling through woods in search of food, and keeping together by uttering

▼ *Many species of songbird, such as these Starlings, roost together for safety and warmth*

▲ *Starlings also gather in spectacular pre-roost displays at dusk on winter evenings*

brief contact calls. Flocks like this have been shown to be more efficient in finding food and avoiding predation, thus making it a benefit for each individual bird to join. Seed-eaters such as **finches**, **sparrows** and **buntings** do the same.

Waders and **wildfowl** also form large flocks outside the breeding season. In most cases this is because their food is very plentiful in specific areas such as mudflats or estuaries, but not found elsewhere. Flocking has the added advantage of making each bird less vulnerable to predators such as birds of prey, which may be confused by the swirling mass of birds, and unable to focus on a particular individual. It only takes one bird to spot a hunting **Peregrine**, and give an alarm call, and the whole flock can take to the air, wheeling from side to side in an effort to confuse the attacker.

A few species only form large flocks occasionally, such as a feeding frenzy of **gulls** behind a trawler, or a gathering of **Magpies** in an area with a high population of the species. Again, to do so must confer an advantage, however small, on each individual – otherwise they will disperse and search for food on their own.

The avian habit of forming flocks can be a great help to birdwatchers, especially in woodland or farmland habitats. At first, it can sometimes seem as if there are no birds present at all. But watch for movement, and listen for the contact calls made by members of a flock, and sooner or later you'll come across the flock itself.

To observe flocking at its most spectacular, you will need to visit certain specific habitats, such as estuaries, traditionally managed farmland, or a specific site for a **goose** or **Starling** roost. Make sure you check details such as tide tables or the time the sun sets, as these can be critical: arrive too early and the birds will still be out feeding in the marshes or fields; arrive too late and you will have missed the spectacle.

Roosting and sleeping

Flocking is often the prelude to roosting: either for the night, as in most groups such as **gulls**, **crows** and **Starlings**; or because the high tide has temporarily covered up the birds' food supply, as in the case of **waders**. Roosting confers many of the same advantages as flocking: notably warmth during cold nights, and safety in numbers against predators.

The majority of the world's birds follow a typical circadian cycle, linked to the 24-hour rhythm of the day. Thus as evening approaches, diurnal species like the **Starling** leave their feeding areas and gather together at a single roosting place, where they will spend the hours of darkness. Starling roosts have been known to contain several million birds, making them an awe-inspiring sight, especially as they wheel about the sky in vast flocks just before and after sunset. Whether or not this behaviour has any hidden purpose, such as communicating the whereabouts of food resources, is not clear.

Communal roosting like this takes place for several reasons. The first of these is safety: by joining together with others, an individual bird has a far greater chance of avoiding predators. A second, particularly during the winter months, is warmth – by huddling together, birds are better able to conserve valuable body heat.

▼ *Jackdaws and Rooks often gather in winter evenings at large, noisy roosts*

◀ *Pied Wagtails gather in winter roosts in towns to keep warm and avoid predators*

Birds that do not usually form flocks during daylight hours often roost together: for example, **Pied Wagtails**, which tend to feed singly or in pairs during the day, come together in roosts of 50 or more birds at night. **Wrens**, normally pugnacious and solitary birds, will also gather together, especially during very cold winter weather, when several dozen birds have been found huddling together in a single nestbox!

A characteristic prelude to roosting is a flight line, in which hundreds or even tens of thousands of individuals can be watched as they make their way from their daytime feeding areas to their roosting site. This is especially prevalent amongst **gulls**, which can be seen following strictly defined routes each morning and evening as they travel between roosting and feeding areas, sometimes a distance of many miles. Other birds that regularly follow flight lines include **crows** and **thrushes**.

Not all birds roost communally in large groups. Many songbirds choose to spend the night either on their own or in a loose flock, though the same factors of safety, security and warmth are just as important.

▶ *Long-eared Owls are one of our most elusive resident species, but can sometimes be found at their daytime roost*

A small number of birds, such as **owls** and **nightjars**, are predominately nocturnal, and so roost by day. Owls tend to roost singly, and can sometimes be hard to locate. However, they are creatures of habit, and will often occupy the same roosting place from day to day and even year to year, providing they are not disturbed, making them relatively easy to find once you know a regular site.

Once at a roost, birds often seem so noisy that you wonder how they can ever get any sleep. This may be because the human notion of eight hours of uninterrupted, deep sleep is quite different from the experience of most birds. Even when safely together in a roost, birds always need to be wary of potential predators, so they tend to 'cat-nap', often tucking their head under their wing, and sleeping while standing up. Some birds, notably **Swifts**, appear to be able to sleep on the wing, but due to the difficulty of actually observing this phenomenon it has not been shown how they do so.

Not all birds sleep at night. Those dependent on the state of the tides for their feeding, especially waders such as **sandpipers** and **plovers**, must attune their bodily rhythms to tidal rather than diurnal cycles, and can often be seen fast asleep during broad daylight. Conversely they, and other species such as **ducks**, often feed by night, especially if there is a full moon to enable them to find their prey more easily.

Many coastal species of waders, such as the **Dunlin**, **Oystercatcher** and **Knot**, are only able to feed when the tide is out, exposing the food-rich mudflats of estuaries and saltmarshes. So as high tide approaches, they need to find a suitable place to roost. Being in the right place at the right time when thousands of waders come to roost can be an unforgettable experience. Two or three hours before high tide, as water starts to cover their feeding areas, they begin to form large flocks. This is often the best time to observe them, as they wheel around the sky or move from place to place.

Then, as the tide finally reaches its height, they gather together in vast numbers, either occupying the few small areas of mud left uncovered by the sea, or roosting on nearby non-tidal areas such as gravel pits. Once settled, they rest and sleep for several hours, before the waters recede and they are able to feed once again.

▼ *Many waders such as Dunlin sleep during the day when high tides mean they are unable to feed*

Chapter 2
Feeding

Food types and feeding methods

One of the ways in which birds have adapted particularly widely is in their feeding methods. Between them, the world's birds eat almost anything: from shellfish to seeds, berries to bees, and nuts to nectar. Some even eat other birds! So it is hardly surprising that the various species and families have adapted to feed in many different ways, and in doing so have developed an extraordinary range of differently shaped bills.

Even within a single group, such as the songbirds, there is a vast array of different feeding methods. Seed-eaters such as **finches**, **sparrows** and **buntings** may prise their food from the heads of fruiting plants, or simply pick up seeds from the ground. Some, like the **House Sparrow**, have all-purpose bills designed to enable them to take a wide variety of different seeds; while others, like the **Goldfinch**, have highly specialised long, slim bills which have evolved to obtain seeds from plants such as the teasel, which other finches find it impossible to get

▼ *Goldfinches have a needle-sharp bill, perfect for prising the seeds out of teasels*

to. The most specialised seed-eaters of all, **crossbills**, have evolved an extraordinary and unique bill in which the tips of the mandibles cross over, enabling them to prise open pine cones and obtain the seeds.

Insect-eaters, too, have evolved a wide range of feeding methods. Many **warblers** obtain their food by gleaning: working their way around the leaves, twigs and branches of a tree, picking up tiny insects as they go. **Goldcrests** have developed this into a method known as 'hover-gleaning', in which they flutter like a hummingbird while seizing tiny morsels from the tips of leaves.

Flycatchers, as their name suggests, launch themselves into mid-air, grab a flying insect, then return to their perch to devour it. **Bee-eaters** do the same, knocking the bee to stun or kill it and remove the sting before consuming it whole. Meanwhile flying hunters such as **Swallows**, **martins** and **Swifts** have broad, wide mouths which enable them to engulf flying insects in mid-air.

Other songbirds are omnivorous, and have more generalised bill shapes and feeding methods. **Tits** can take seeds and insects, and mainly feed their young on caterpillars. Members of the **crow** family such as the **Jay** and **Magpie** will eat a very wide variety of foods, and have all-purpose bills to do so. **Thrushes** also take a variety of food, but have fairly long bills enabling them to catch earthworms, while they are also able to eat berries in winter. The ability to vary diet from season to season is a crucial factor in survival, especially for those birds that, although insect-eaters in spring and summer, stay put in northern latitudes for the winter. A good example is the **Blackcap**, which in winter will change its insectivorous diet to one of seeds and berries, enabling it to overwinter in Britain.

▲ *Blackbirds love to feast on juicy berries, which provide plenty of energy*

◀ *Spotted Flycatchers launch themselves from a favourite perch to catch flying insects*

Many of these more versatile species have adapted very readily to food provided in artificial feeders by us. Once, this habit was confined to **tits** and **sparrows** on hanging feeders, and **Starlings**, **Robins** and a few other species on bird tables. But a revolution in the design of bird feeders and the quality and variety of food has enabled many more species to feed in this way. Today it is a common sight to see quite specialised feeders such as **woodpeckers** on bird feeders.

Wildfowl, too, show a range of different feeding methods. Many species of **swans**, **geese**, and **ducks** such as **Wigeon** regularly graze on grass or other crops such as sugar beet (often to the annoyance of farmers). Others, especially the so-called 'dabbling ducks' such as **Mallard**, **Teal** and **Pintail**, dabble for morsels of food on the surface of water, while the **Shoveler** has a feeding method all its own. **Swans** dip their long necks below the water surface, while 'diving ducks' such as **Tufted Duck** and **Pochard** dive down beneath the surface to find their food.

More than any other group of birds, **waders** show great adaptability when it comes to feeding methods, as shown by their great variety of bill sizes and shapes. Many, including common species such as **Sanderling** and **Redshank**, have fairly straight, longish bills, to probe beneath the mud or pick items off the surface. Others, including the **Curlew** and **Avocet**, have longer, curved bills, used to probe for lugworms in the case of the Curlew or sift the water for tiny invertebrates in Avocets. Watching a variety of waders feeding is an object lesson in how evolution produces different body shapes from the same basic design.

Seabirds have also evolved a wide variety of ways to obtain food. Some, like the **Gannet** and many species of **terns**, fly over the water, then plunge down to obtain food. Others, such as **auks**, swim on the surface, and then dive down, sometimes hundreds of metres below. **Gulls** are generalists, either grabbing morsels from the surface or leaving the sea altogether to feed in fields or even rubbish dumps, where they feast on our household food waste.

▼ *Unlike most ducks, Wigeon often graze on grassy wet meadows*

▲ *Knots have an 'all-purpose' bill which they use to feed in mud or shallow water*

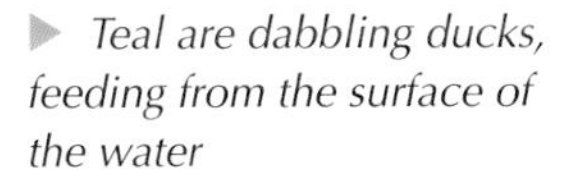

▶ *Teal are dabbling ducks, feeding from the surface of the water*

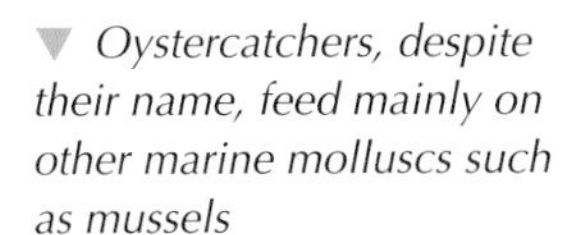

▼ *Oystercatchers, despite their name, feed mainly on other marine molluscs such as mussels*

Predators

In a sense, any bird that feeds on any other living creature is a predator. However, in general we reserve the term 'birds of prey' for two unrelated but superficially similar groups: **day-flying raptors** and **owls**.

The first of these groups includes a wide range of different species with an equally wide range of hunting and feeding methods: **eagles**, **hawks**, **Common Buzzard**, **harriers** and **Red Kite**, **falcons** (which although superficially similar to hawks are actually in a quite different taxonomic order), and the fish-eating **Osprey**. They all share the characteristic traits of fierce talons to catch and grip their prey, and a sharp, hooked beak to tear it apart for feeding. However, they hunt and catch their victims in a wide range of different ways.

One of our commonest and best-known birds of prey, the **Kestrel**, often hunts by hovering motionless over a grassy verge, then plunging down to catch an unsuspecting vole. In contrast, the **Hobby** grabs insects such as dragonflies, or even small birds like martins, in mid-air, while the **Peregrine** hunts by 'stooping' from a great height down onto its prey.

Sparrowhawks use their short wings and long tail to manoeuvre their way through dense woodland or garden foliage, ambushing their target; while larger raptors such as **eagles** and **buzzards** fly much higher above their territory, searching for food. The **Osprey** hunts in a unique way for a raptor: plunging feet first into a lake to grab large fish in its talons.

Owls are mainly nocturnal, and tend to hunt by watching and waiting from a tree or post (**Tawny** and **Little Owls**), or gliding low over open ground (**Barn Owl**).

▼ *One folk-name for the Kestrel is 'windhover', from its habit of hovering in mid-air to hunt its prey*

Little Owls feed on a range of prey including small mammals, beetles and earthworms

Specialist feeders

A few species have evolved highly specialised feeding methods: either to exploit a particular kind of prey (for example **Bee-eaters** and bees), or to gain access to a food resource that would otherwise be denied to them. So **Kingfishers** sit on perches above streams and ponds, then plunge below the surface to grab small fish, behaviour not shared with any other British bird.

Sometimes a bird may adopt feeding methods quite different from its relatives. For example, another aquatic bird, the **Dipper**, is related to the thrushes, yet it has evolved to exploit a very different environment from their terrestrial one. Dippers live on fast-flowing streams and rivers, and perch on rocks before plunging right beneath the surface of the water to search for small aquatic invertebrates.

Perhaps the most bizarre, yet highly effective, feeding method of all is known as kleptoparasitism, or, as it is often called, piratical behaviour. It is commonly practised by **skuas**, some species of **gulls** and sometimes birds of prey, and involves one or more birds chasing an unfortunate victim (usually of another species such as a small gull or tern), and forcing it to either drop, or in some cases regurgitate, its food, which the chasing bird then seizes and swallows.

▲ *Kingfishers are the ultimate river hunter: plunging into the flowing waters to grab an unsuspecting fish*

▼ *Great Skuas – also known as 'bonxies' – are kleptoparasites, chasing other seabirds to steal their food*

This behaviour is often observed in seabird colonies, where species such as the **Arctic Skua** (or 'Parasitic Jaeger' as it is known in North America) make the lives of terns very difficult indeed. A similar opportunistic form of piracy is practised by many species of the **crow** family, especially **Magpies**, which will take the opportunity to seize food from other birds, or even from mammals such as Foxes, if they get the chance.

Drinking

It is easily forgotten that as well as feeding, birds also have to drink. In fact some species, especially insect-eaters, obtain much of the moisture they require from their food, and may only have to drink every one or two days. However, seed-eaters such as **finches** and **sparrows** have to compensate for the lack of moisture in their food by drinking at regular intervals. They do so either by visiting streams, ponds or puddles, or by coming to artificial sources of water such as garden ponds and birdbaths.

Birds are unable to swallow in the way we can, so need to drink in particular ways. These include sipping a drop of water into the bill, then tilting their head back to allow the liquid to pass down their throat; others do a form of sucking which involves vibrating the tongue to pump water back into the gullet. Drinking birds are very vulnerable to attack by a predator, so they are often extremely wary, taking only a sip or two before looking round to make sure they are still safe.

A few species, notably **Swallows**, drink on the wing, swooping down to the water and skimming the surface as they go. This has the added advantage of reducing the risk of being caught by a predator.

Seabirds, especially ocean-going species such as **shearwaters** and **petrels**, have another problem: how to obtain water while at sea. They have evolved a way to drink seawater safely, using salt glands to remove the excess salt. This enables them to stay away from land for many weeks or even months on end.

▼ *Willow Warblers, like other small birds, need to drink and bathe regularly*

Chapter 3

Breeding

Timing

In spring, the thoughts of young men turn to love, along with the rest of the animal kingdom – and birds are no exception. During a brief few months at the start of the year, they must make hay while the sun shines: or to put it more specifically, they must breed.

This is a complicated business which in general goes like this: the male bird must stake out a territory, fight off rival males, and find a partner (or partners). The pair then engage in complex courtship rituals, build a nest, and mate; the female then lays her eggs and incubates them (sometimes with the male's help). Once they hatch the real work begins, as the young constantly demand to be fed. Finally, the young fledge and leave the nest, though even then they may require parental care. No wonder birds look tatty and exhausted by the time midsummer comes!

▼ *Mute Swans are usually faithful to their partners for life*

Not all birds breed at the same time of year, by any means. Resident birds such as the **Blue Tit** and **Blackbird** habitually breed a month or more earlier than migrants like the **Willow Warbler** or **House Martin**, though this is not always the case. Some migrants, such as the **Chiffchaff**, arrive back in March and get down to breeding straight away, while residents like the **Yellowhammer** may not start until June. **Crossbills** may lay eggs in January, while the first brood of **Mallard** ducklings often appears well before Easter. On the other hand some species, especially **pigeons** and **doves**, appear to breed virtually all year round!

Some species may have several broods a year (like the **Blackbird**, which can have up to five, and may be nesting virtually continuously from March to August!). Others may have failed to breed at their first attempt, and try again much later – I once saw a brood of **Great Crested Grebe** chicks hatch in mid-September, and still survive the winter unscathed.

In recent years timing of the start of the breeding season has changed. Climate change means that the start of spring comes a week or more earlier to many parts of Britain, and the birds have reacted accordingly, laying eggs up to two weeks earlier than 30 years ago. Indeed, the discovery of this phenomenon, using data collected by generations of amateur birders, was one of the first pieces of empirical evidence that global climate change was real rather than hypothetical.

With the recent onset of mild winters, some common species may even be tempted to begin nest-building before Christmas, though a cold spell will soon put a stop to this.

◀ *This fledgling Blackbird has just left the nest, and is still dependent on its parents for food*

Territory and song

A spell of mild weather in February, and our woods and gardens are suddenly filled with the sounds of spring. **Song Thrushes** sing their repetitive tune from the tops of roofs; **Great Tits** bounce around the blossoming bushes, calling 'tea-cher, tea-cher'; and deep in the undergrowth, tiny **Wrens** explode in a frenzy of song.

Even the arrival of late-winter snow and ice may only bring a temporary halt to the chorus. Whatever appearances suggest, spring is here, at least for the birds. But to paraphrase the 1950s pop song, why do birds sing in the first place?

Broadly speaking, birdsong has two main functions: to defend a territory and attract a mate. Male birds generally arrive back on their breeding grounds earlier than their mates, and spend the first few days moving around their new territory, establishing the boundaries and advertising their presence to all-comers – especially rival males of the same species.

Some species, notably the **Robin**, also sing during the autumn and winter, as unlike most songbirds they also hold winter territories. On a cold, dull day theirs may be the only sound to be heard, but on a bright, sunny day in February all kinds of other singers join the chorus, in anticipation of the breeding season to come.

Birdsong has a second, equally important function: to attract a mate. As the females leave their winter quarters and return to their breeding habitats, males are in desperate competition, for those that fail to breed run the risk of dying before they can pass on their genes. This is especially true for many songbirds, whose life expectancy is only a year or so.

▼ *Robins sing virtually the whole year round, as they hold territories in autumn and winter as well as spring and summer*

One noticeable thing about birdsong is that in general, the best singers have the dullest plumages. The **Robin** is a notable exception, but how about the **Blackbird**, **Nightingale** and **Marsh Warbler**? The explanation is obvious when you think about it: birds with brightly coloured plumages don't need to sing a complex song to attract a mate; those that are dull brown or black do! And of course the opposite applies: birds with bright plumages such as the **Bee-eater** or **Kingfisher** have no need for a complex, melodious song.

Along with food and habitat, birdsong is one of the most important factors in the ecological isolation of different species. This was first discovered by the great 18th-century naturalist Gilbert White, who managed to distinguish between the three species of 'willow-wrens' (**Chiffchaff**, **Willow Warbler** and **Wood Warbler**) by listening to their distinctive songs.

Courtship, display and mating

Once a singing male has attracted a mate, the serious business of courtship and pairing gets underway. After feeding, courtship is probably the most important aspect of bird behaviour. It takes all kinds of forms, and far from being confined to what we think of as spring, can begin as early as January.

Many species pair up during the winter when they are in flocks, while others, such as the **Mute Swan**, form lifelong pair bonds, which usually only end when one of the partners dies. Other species, such as the **Osprey**, form an attachment to a nest; and as a result often breed with the same partner from the year before.

Great Crested Grebes have one of the most spectacular of all courtship displays. The two birds face each other in the water like a pair of teenage lovers, rubbing their bills together and shaking their heads in a ritualised pantomime to cement the pair bond.

If you're really lucky, you may even see them go on to perform the memorable 'Penguin Dance', during which both birds gather water weed in their bills, then appear to stand up in the water, frantically paddling with their legs while waving the weed in each other's faces. Whatever turns them on, I suppose...

One of the easiest courtship rituals to observe is that of the **Feral Pigeon**. The male puffs himself up like a prizefighter, then performs a little dance around the female, who generally looks singularly unimpressed by all the fuss. There may then be a quieter period during which the male feeds the female, or they preen one another. Only then, once the pair bond is fully formed, will the male will attempt to copulate with the female, though as often as not she will foil his advances.

A stunningly beautiful bird, the **Avocet**, performs another wonderful display. The male approaches the female tentatively, dipping his decurved bill into the water as if attempting to feed. Then, when he judges that she is receptive to his advances, he leaps onto her back, mates in a second or two, and jumps off. This is followed by an extraordinary little ritual, in which he runs forward away from her pecking at the water, while she carries on feeding.

▼ *Feral Pigeons, like other species of pigeon and dove, have complex courtship displays*

▲ *Avocets pair up in spring, bowing to one another to cement the pair bond*

◀ *Great Crested Grebes have one of the most complex courtship displays of any British bird, known as the 'penguin dance'*

All this palaver has a serious purpose, of course. The male who can most impress the female is the one who will get the chance to mate and reproduce. And the female is under pressure too. She must choose the healthiest-looking male, to increase her chances of producing a long line of descendants. Courtship rituals may look like a bit of fun, but to the birds themselves, they truly are a matter of life or death.

Nest-building

Once the courtship ritual is over, and the pair bond is strong, nest-building usually begins. To say 'building' is actually not always accurate, as many species simply deposit their egg or eggs on a suitable level surface. This is especially true of colonial species such as seabirds: a **Guillemot** lays its egg in a depression on a cliff shelf; the egg itself is pear-shaped, so if accidentally nudged it rolls in a very tight circle and does not fall off. Other birds such as **Stock Doves**, most **tits** and **Starlings** lay their eggs inside a hole in a tree, requiring the minimum amount of 'building' effort.

But for most birds a nest is required in order to keep the eggs safe and allow them to be incubated. Nests come in a whole range of shapes and sizes, and also vary in the amount of care and attention taken to make them.

The 'classic' songbird nest is that made by species such as the **Blackbird** or **Song Thrush**: a neat cup of woven grass or small twigs

▲ *Puffins nest in old rabbit burrows, where their single egg and chick can be safe from predators*

lined with mud. Larger birds, such as the **Wood Pigeon**, make a much tattier-looking nest from twigs, which often looks so flimsy you can see the eggs through it from beneath! One of the most ornate and complex structures is that built by the **Long-tailed Tit**: a ball of hair, moss and feathers (up to 2,000 in a single nest), held together by lichen. It is this amazing nest that gave the species the old country name of 'bumbarrel'.

Other small birds, such as **warblers**, often nest on the ground, or build their nest in the fork of a tree. **Goldcrests** hang their tiny nest and its precious contents on the end of a twig, the structure being so light it manages to stay put.

Waterbirds such as **Coots** and **grebes** build floating nests out of aquatic vegetation; in the case of the **Great Crested Grebe**, once the eggs are laid the adults use vegetation to cover them up when they leave the nest to search for food.

The largest nest of any British bird belongs to the **Golden Eagle**: a bulky structure of twigs up to 3 metres across and 5 metres deep. One of the smallest is that made by the **Wren**, but there is a catch: the male must make several nests before the fussy female is satisfied and chooses the best one in which to lay her eggs.

▲ *Grey Herons are one of the earliest birds to begin breeding, in noisy colonies known as heronries*

▶ *The Long-tailed Tit builds one of the most complex of all bird nests, made from feathers, lichen and spider's webs*

Egg laying and incubation

Once the nest is built, the female can lay her eggs: anything from one (most **seabirds**), three to five (many **waders**) or four to 12 (**songbirds** and **wildfowl**) to 20 or more (**gamebirds** such as the **Pheasant**). These are generally laid one per day, with incubation usually beginning the day after the final egg is laid and the clutch is complete.

This is a risky time for the birds, as there are many predators for whom a clutch of eggs makes a tasty meal, such as **Jays**, **Magpies** and **squirrels**. The weather can play a part too, especially a cold snap, or heavy rain, which can cool the eggs and kill the embryos inside. So incubation is a full-time job for most birds: it is either carried out by the female alone (as in many lekking species such as the **Ruff**, **Capercaillie** and **Black Grouse**), the female with the help of the male bringing food (most **songbirds**), or by both sexes (most **seabirds**). In two rare British breeding species, the **Dotterel** and the **Red-necked Phalarope**, the roles are completely reversed, and it is the male who incubates the eggs while the female goes off on her own (see p. 44).

The period of incubation varies greatly: from just 11 days for the **Lesser Whitethroat** and **Skylark**, to an astonishing 54 days in the case of the **Manx Shearwater**, and even longer for some **Fulmars**. For most songbirds the usual incubation period is around two weeks; ducks, waders and gamebirds about three to four weeks; birds of prey, four to seven weeks; while seabirds have the longest incubation periods, of

▼ *Carrion Crows are famous for stealing eggs from other birds' nests*

▲ *Like many seabirds, Fulmars only lay a single egg, which they incubate for up to eight weeks*

between four and almost eight weeks. This may be because the young need to be born with fat deposits so that they can go several days without food when their parents are away at sea.

The timing of incubation is usually fixed so that the young hatch out within a few hours of one another, meaning their fledging date will also be approximately the same. Exceptions include the larger birds of prey such as **eagles**, which lay two eggs but incubate the first one immediately, so that it hatches a day or two before the second.

Parental care and fledging

Once the young have hatched, the hard work really begins. Young birds can be divided into two very distinct groups: those that are born blind and remain in the nest for several weeks, requiring constant care and feeding from their parents; and those that leave the nest straight away, able to walk and in some cases also swim, and find food for themselves or with the help of their parents. The first group, which includes all the passerine birds (or **songbirds**) are known as altricial, or nidicolous, species; while the second group, which includes **wildfowl**, **waders** and **gamebirds**, are known as precocial (or nidifugous) species. Some groups of birds, such as **gulls** and **terns**, are known as semi-precocial,

▲ *These Swallows are about to leave the nest, which looks as if it will soon be too small to accommodate them*

as they are born able to see and with a downy coat, but stay in or very near the nest and are fed by their parents for some days after hatching. When the chicks hatch it is crucial that they know who their parents are, so they can follow them as they search for food. So the chick 'imprints' on the first living creature it sees – usually its mother or father.

Interestingly, all these strategies work very well; indeed, for them to be adopted by such a wide range of different species they must be advantageous in an evolutionary sense. For **songbirds**, which tend to hide their nest away in foliage or holes in trees, the altricial strategy means that their young are kept safe from predators, while they collect food in the immediate area of the nest. For aquatic birds such as **ducks**, and other precocial species such as **gamebirds**, all of which nest on the ground or on water, the strategy is equally sensible, enabling the young to range far and wide in search of food while avoiding predators by swimming, running or hiding.

Fledging is a term that technically only applies to altricial birds, and refers to the period when the baby birds have acquired their first feathers (juvenile plumage), and are able to leave the nest and fly. Even at this time most baby songbirds are highly dependent on their parents for food and safety, and despite the adults' best attentions this is the time when the death rate is the highest.

Fledging is also, however, used to describe a similar process in precocial species: this time not the point at which they leave the nest (which is immediately or a few hours after hatching), but the point at which they lose their downy plumage, acquire their first proper feathers and are able to fly.

◀ *Arctic Terns often attack visitors who venture too close to their nests, eggs and chicks*

Like incubation periods, the time taken to fledge varies enormously from species to species and group to group. Most young songbirds fledge roughly two to three weeks after hatching, though this period can range from just 11 days for the **Lesser Whitethroat** and 12 days for the **Dunnock**, to 24 days for the **Swallow**. Crows stay in the nest even longer, with young **Jackdaws** and **Carrion Crows** taking up to five weeks to leave.

For larger birds the fledging period can be much longer. Once again, **seabirds** hold the record, with young **Manx Shearwaters** taking up to ten and a half weeks. The young ultimately must leave the burrow or starve, as the parents stop feeding it a few days before it reaches fledging age.

Hybridisation

One of the most puzzling things facing a relatively inexperienced birder is coming across a bird which, although it doesn't match any species in the field guide, appears to have some of the characteristics of two other species. Sometimes these are birds with plumage aberrations, such as leucistic or melanistic birds. But in other cases they may be hybrids.

The extent of hybridisation varies dramatically between different species and groups. In some families, including many **songbirds**, it is virtually unknown; while in others, notably **wildfowl**, it is very common indeed. For example, there are many recorded instances of

▲ *Ruddy Ducks are now being eradicated from Britain as this non-native species was causing problems by hybridising with a rare European relative, the White-headed Duck*

hybrid **ducks** and **geese**, some of which have caused confusion amongst even very experienced birders, as they tend to resemble a third species rather than either of their parents (as in **Tufted Duck** x **Pochard** hybrids which look like a rare American vagrant, the **Lesser Scaup**). **Geese** are particularly prone to hybridisation, both between two wild species, and one wild and one domestic bird. Birds of prey rarely hybridise in the wild, but misguided falconers often artificially pair species, including such bizarre couplings as a **Merlin** x **Peregrine**.

Hybridisation may appear to be a futile gesture, as the offspring tend to be infertile (indeed, the definition of a 'proper' species used to be that if it hybridises with a close relative the young are not fertile). However, in a few cases the two parent species share enough genes to result in a stable hybrid population being formed, which may ultimately evolve into a new race or species.

Perhaps the best-known recent case of large-scale hybridisation is between the North American **Ruddy Duck** and the rare **White-headed Duck** in southern Spain. Escapee Ruddy Ducks from exotic wildfowl collections in Britain prospered in the wild and spread southwards across Europe, potentially threatening the entire south-west European population of the White-headed Duck through hybridisation. As a result, the feral Ruddy Ducks are in the process of being eradicated from Britain and Europe to prevent the loss of their sibling species.

▲ *This young Cuckoo, being fed by its Reed Warbler foster-parent, is now almost too big to fit into its nest*

Unusual breeding behaviour

A few species reverse the usual male/female roles when breeding. In Britain this is practised by two scarce breeding waders, the **Red-necked Phalarope** and **Dotterel**. In these species the sexual roles are reversed to such an extent that it is the female who sports the brighter plumage and takes the lead in courtship. In common with some other waders, the female then lays her eggs and leaves all the duties of incubation and care of the young to the male, while she seeks a second male. In some cases female Dotterels have been recorded mating first in Scotland, then leaving the male to incubate the eggs while they fly to Scandinavia to mate once more and raise a second brood.

Perhaps the best-known form of 'aberrant' breeding behaviour is the practice of laying eggs in other birds' nests, known as brood parasitism. This is found in several of the world's bird families, but only one species does it in Britain: the **Cuckoo**.

Cuckoos arrive back from their African winter quarters in May, and immediately set about the task of mating and laying eggs. But unlike conventional species the female Cuckoo does not build a nest herself, but lays her egg in that of a host species (almost always the same species in whose nest she was herself raised). The commonest hosts in Britain are the **Reed Warbler**, **Meadow Pipit** and **Dunnock**.

First, she removes an egg already laid by the host, before depositing her own (which is usually patterned to match the host's egg) with great speed. Once hatched, the young Cuckoo instinctively ejects any remaining eggs and/or chicks, gaining a monopoly of its parents' attention. The Cuckoo and its hosts are engaged in a constant form of escalating warfare, which the Cuckoo appears to be winning: by this method the female can lay up to 20 eggs, and still leave for Africa well before her offspring have even fledged.

Polygamy

Polygamy (having multiple partners) is practised by many different species. The male **Corn Bunting**, for example, practises polygyny, sometimes having seven or more females nesting within his extended territory. In waders such as the **Sanderling**, the females practise polyandry, pairing sequentially with more than one male. It used to be thought that while only a few birds (such as **Mute Swans**) pair more or less for life, most are at least seasonally monogamous, the pair staying faithful to each other for the duration of a full breeding season.

However, scientists have recently discovered that many birds that appear to form monogamous pair bonds are not so faithful to their mates as was once thought. For example, the male **Dunnock** jealously guards his spouse, who will often mate with any other single males in the immediate area. Male Dunnocks have even resorted to tugging the sperm out of the female's cloaca, before re-mating to ensure that it is their offspring that are hatched.

In other cases females will take a single partner to share nest-building and incubation duties, but still try to mate with as many males as possible. The idea is that by doing so the female maximises her chances of producing the healthiest and fittest young, and passing on her genes to future generations. Males, however, must guard their mate to ensure that they do not become cuckolds to someone else's young. For even where a pair appears to be monogamous, DNA studies have revealed that the eggs in a single clutch may have different fathers, due to the female mating clandestinely with other males.

▶ *Corn Bunting males are polygynous, and may have as many as seven different females at the same time*

▼ *Dunnocks practise a complex range of sexual behaviours, with liaisons outside the main pair bond very common*

Chapter 4

Migration and navigation

Why do birds migrate?

It has been estimated that as many as five *billion* birds, of more than 200 different species, undertake the twice-yearly journey between Eurasia and Africa. These include such diverse groups as **waders** and **warblers**, **swallows** and **terns**, proving beyond doubt that as an evolutionary strategy, migration is a success.

But why do birds migrate? What is the point of leaving a comfortable home in Britain, and heading thousands of miles south, facing all kinds of hazards along the way? And if a bird manages to survive this perilous journey, and reach its winter home, what impels it to come back here the following spring? Why not just stay put the whole year round?

The traditional reason given is that these birds are unable to survive the northern winter, and must head south to find food. This is true, as far as it goes, but it is only half the story.

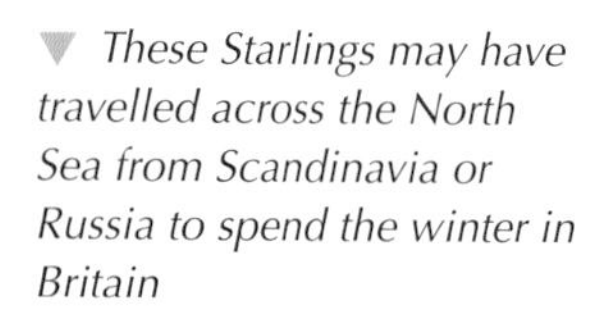

▼ *These Starlings may have travelled across the North Sea from Scandinavia or Russia to spend the winter in Britain*

In fact to view migratory birds as 'our' birds heading south for the winter is to put the cart before the horse. Species like **swallows**, **warblers** and **flycatchers** originated in Africa, and first headed north to avoid competition with other species there. By travelling northwards, they found 'spare' ecological niches where they could breed and raise their young, with the advantages of abundant food, long hours of daylight, and fewer competitors.

▲ *Although it weighs barely an ounce, this Swallow will fly thousands of miles to Africa and back*

▼ *Sedge Warblers make the journey to West Africa in one or two giant leaps each autumn*

▲ *British Blackbirds stay put for the winter – but is that a riskier strategy than migrating?*

But at the start of autumn, as the temperature begins to drop, the insects on which so many of these species rely for food begin to disappear. Meanwhile, the days get shorter, so there is less and less time available to forage for scarcer and scarcer resources. So for many insect-eaters there are only two choices: either stay put and starve to death, or head south for the winter. To survive, they developed adaptations including longer wings, a range of navigational devices, and the capacity to store large amounts of fat for the migratory journey.

We often assume that migrants face far greater dangers than resident species, but in fact the opposite may be true. Migration expert Peter Berthold has pointed out that while summer migrants such as the **Sedge Warbler** tend to produce a single brood of young, resident species like the **Wren** or **Blackbird** raise two or more broods, some species laying more than a dozen eggs at a time. This would suggest that staying put during the northern winter leads to a higher mortality rate than the journey back and forth to Africa.

The Swedish scientist Thomas Alerstam has gone even further, turning the question on its head. He asks why, given the obvious advantages of migration, any birds pursue the resident strategy at all. As he puts it: 'Why do not *all* birds migrate?'

How do birds navigate?

Ever since the Ancient Greeks tied messages to the legs of migrating birds, in the hope of discovering where they spent the winter, man has been fascinated by bird migration. Even now, when you look at a **Willow Warbler** or **House Martin**, it is hard to believe that such a tiny bird is capable of travelling such great distances across the globe.

▲ *Millions of House Martins migrate to Africa, and yet we have very little idea exactly where they spend the winter*

Indeed, until relatively recently the prevailing theory suggested that like many mammals, birds spent the winter months in hibernation. Sightings of **Swallows** gathering over water in autumn reinforced this theory, with observers claiming to see the birds plunge beneath the surface of water. Even the great 18th-century ornithologist Gilbert White subscribed to this theory, at least in part, though he did have his doubts.

Yet as long ago as the days of the Old Testament people suspected the truth: that birds travel south in autumn to avoid the colder months. The best evidence for this early belief in migration is from the *Book of Jeremiah*: 'Yea, the stork in the heavens knoweth her appointed times; and the turtle-dove and the swallow and the crane observe the time of their coming.'

The concept that birds 'know the time of their coming' is surprisingly close to the truth. Many birds time their journeys by changes in day-length, though local factors such as weather conditions also dictate timing of departure. Day-length affects the birds' endocrine system, producing hormones that stimulate them to prepare for the long journey ahead, for example by putting on extra fat supplies. Once they have returned, the same glands are responsible for prompting the onset of the breeding cycle.

Once on their way, birds use a complex hierarchy of cues to enable them to find their way, rather than relying on a single navigational method. These include navigating by the stars and sun, as the old sailors used to do, using polarised light, and of course the earth's magnetic field, which all enable them to head in the right direction. Once they get much closer to home, they rely on more visual cues such as local topography, often following coastlines or rivers to steer them to their destination.

▲ *Raptors such as this Montagu's Harrier migrate by day*

None of these methods is infallible. Some birds, particularly juveniles undertaking their first migratory journey, appear to have something wrong with their internal 'clock', and may head in entirely the wrong direction. Others are diverted off course by unusual weather conditions, especially cloud and rain, which make it impossible to use visual cues such as the stars or topography. Some groups of birds, especially migrants crossing large areas of water, appear to be 'drifted' off course by crosswinds, though they are usually able to reorientate themselves successfully.

Migration strategies

Not all birds migrate in the same way, or using similar strategies. Some, like **warblers** and **chats**, migrate by night, spending the day feeding or sheltering from predators. Others, like **Swallows** and **raptors**, travel by day, feeding as they go.

Some, like the **Knot** and **Sedge Warbler**, choose a 'long-haul' strategy, putting on vast amounts of fat before completing their journey in a few huge leaps of hundreds or even thousands of miles at a time. Others, like the **Swallow**, potter along from place to place and take several weeks or even months to complete their journey.

Across the Atlantic Ocean, the **North American warblers** choose to fly across wide expanses of sea in order to shorten their journey; whereas other species, especially large birds of prey such as **eagles** and **buzzards**, try to avoid crossing water at all, instead concentrating in huge numbers at land crossings such as Gibraltar, Eilat in southern Israel, and the Bosphorus in Turkey.

Each of these strategies has its advantages and disadvantages, but each has evolved over many generations to suit the particular species. It is only in recent years, with ringing studies and the ability to radio track species such as the **Osprey** and **Cuckoo**, that we have begun to understand the subtleties of migration strategy. As technology advances we may soon even be able to track the smallest migrants such as the **House Martin**, to discover where they go when they leave our shores.

Not all species use the same strategy for both the outward (autumn) and return (spring) migrations. For example, those North American warblers which head over the open ocean in autumn follow a quite different route in spring, heading along the coast in short hops. This is because in spring there are no suitable tailwinds to enable them to cross the ocean in the two or three days needed to reach landfall.

This strategy, known as 'loop migration', is also followed by several European species, such as the **Red-backed Shrike**. In autumn, they take a south-easterly route out of Europe, via the eastern Mediterranean, to winter quarters in Central and Southern Africa. In spring, they return by an even more easterly route, across the Arabian Peninsula, probably because meteorological factors make this course more favourable. British **Sand Martins**, too, follow an anti-clockwise route, heading south via Iberia and north-west Africa, but returning via the central Sahara, Italy and central Europe.

◀ *Sand Martins are one of the earliest migrants o return to Britain, usually reaching our shores in March*

Even within a particular species, different populations pursue different strategies. One of the most fascinating of these is known as 'leapfrog migration', and involves a population breeding at a higher latitude 'leapfrogging' over a population of the same species which breeds at a lower latitude. So, for example, Arctic populations of waders such as the **Ringed Plover** migrate all the way to sub-Saharan Africa, while birds breeding in southern Scandinavia only travel as far as southern Europe or North Africa. British-breeding Ringed Plovers follow an even more sedentary strategy, being more or less resident, although some do move westwards to areas of milder winter weather.

Thomas Alerstam has proposed two possible explanations for the existence of leapfrog migration, both to do with competition and timing. The first theory suggests that more southerly birds finish breeding earlier, and move to the nearest available suitable wintering areas. More northerly birds finish breeding later, and by the time they travel south these areas are already 'full', being occupied by the southerly breeders. This forces the northern populations to head farther south still, until they come across somewhere to spend the winter.

The second, more plausible, theory is to do with the timing of breeding itself. Alerstam suggests that it is crucial for birds breeding in temperate zones to winter near their breeding grounds, as the start of their breeding season may vary considerably, being primarily determined by local weather conditions.

▼ *Cranes travel in flocks, with experienced birds guiding the youngsters along their route*

▲ *Whimbrels migrate in small flocks to West Africa and back*

In contrast, Arctic breeders have only a narrow window of opportunity during which conditions are suitable to breed. Therefore they must return to their breeding grounds at more or less the same time every year, whatever the weather conditions. They rely on a sophisticated internal timing mechanism to tell them when to return, so have no need to spend the winter close to their breeding areas.

Unusual migrations

Not all migratory journeys go to plan. Indeed one of the most fascinating areas of birding is observing what happens when things go wrong.

One of the commonest migration-related phenomena is that of 'falls': a sudden and unexpected arrival of migrating birds, usually, though not always, at a coastal 'hotspot'. Falls are generally weather-related, as it usually takes some sort of adverse weather conditions to concentrate the birds and force them to land. For example, during the autumn, migrant songbirds usually depart from Scandinavia during anticylonic weather, which provides clear skies and light following winds to help them on their way. If conditions remain good, they will cross the North Sea, and make their way along the coasts of continental Europe, towards Africa.

But when low-pressure systems are present over the North Sea, migrants often get disoriented, and blown off course by the strong winds. Many fall exhausted into the sea and die. But others carry on and, with luck, make landfall somewhere along the coasts of eastern Britain.

Most songbirds migrate by night, so falls often occur in the early hours of the morning. Some can be spectacular, involving many thousands of birds of many different species. But if you want to experience the thrill of seeing a fall, you have to be quick. Once the birds have rested, fed, and recovered their strength, they're off again, impelled by that mysterious migratory urge.

Another phenomenon occurs in spring, and is known as 'overshooting'. This involves those species that return each year to breed around the shores of the Mediterranean or in France, such as various species of **heron** and **egret**, **Black-winged Stilt**, **Red-rumped Swallow**, **Woodchat Shrike**, **Alpine Swift**, **Hoopoe** and **Serin**.

The key factor in overshooting is the weather conditions, not just here but further south, over continental Europe. Ideally, there should be a large area of high pressure, or anticyclone, situated over the northern shores of the Mediterranean, with its northern limits reaching southern Britain. As the birds arrive at or near their normal breeding grounds, the clear skies and light, southerly winds allow the birds to continue past their destination.

Apart from the meteorological factors, why do birds overshoot at all? One possible explanation is that a tendency to overshoot is genetically programmed, it being a potential advantage for the bird concerned to discover somewhere new to breed. Over the past few years there have been several nesting attempts by overshooting **Black-winged Stilts**, with a pair successfully fledging young in 2014, showing that for some birds at least, the pioneering spirit can lead to success, and may eventually lead this Mediterranean species to colonise Britain permanently.

In a sense, all vagrant birds are 'lost' birds, which will usually perish, but just occasionally they form the pioneering contingent of a new colonist. In recent years former vagrants such as **Cetti's Warbler**, **Mediterranean Gull**, **Great White** and **Little Egrets** have all become regular British breeding species, and others such as **Cattle Egret**, **Glossy Ibis** and **White Stork** look set to follow.

Another form of pioneering spirit can be seen in those few species which are 'irruptive': that is, they occasionally leave their normal breeding and wintering areas to arrive elsewhere in huge numbers. Of these the commonest and most familiar in Britain is the **Common Crossbill**, which regularly turns up in midsummer, and may arrive in large numbers in areas previously devoid of birds. Crossbills generally stay put until the following year, when after breeding early in the New Year they head off again on their nomadic journeys. Another irruptive species, the **Waxwing**, is an autumn and winter visitor here. In some years thousands may arrive; in others hardly any. The size of Waxwing

▶ *Hoopoes are a regular spring visitor to southern Britain, especially during periods of fine weather over mainland Europe*

▼ *Crossbills are an 'irruptive' species, undertaking irregular migrations driven by the search for new sources of food*

irruptions is governed by the availability of berry food on their northern breeding grounds.

Two other species, the **Nutcracker** and **Pallas's Sandgrouse**, are much more occasional irruptive visitors, though when they do occur it can be in huge numbers, to the delight of twitchers.

Not all birds migrate huge distances. An often overlooked phenomenon is known as 'altitudinal migration'. As its name suggests this involves a journey from higher to lower altitudes, usually to spend the winter months where there are more accessible food supplies. Although these journeys may seem insignificant in terms of distance travelled, they often involve a major change of lifestyle. So in winter the **Skylarks** breeding on Britain's upland moors and mountains head to the lowlands, often gathering in flocks near the coast. Even the true 'high-altitude' species such as the **Ptarmigan** will make local movements down the mountain, especially if heavy snow covers their food supplies.

One species makes a doubly unusual migratory journey, moving both downwards to lower altitude, yet also *northwards* for the winter. The **Water Pipit** breeds in the high Alps and Pyrenees, yet a small part of this population spends the winter in southern Britain, usually near fresh water where they can feed on insects.

▼ *Sooty Shearwaters migrate northwards and pass through British offshore waters each autumn*

◀ *Ptarmigan are not strictly migrants, but during hard winter weather they may move down to lower altitudes from their mountain home*

But for perhaps the most dramatic migratory strategy we must turn to three species of seabird, the only ones to reverse the prevailing north–south migratory trend. **Wilson's Storm-petrel** and **Great** and **Sooty Shearwaters** all breed in the southern hemisphere, then head north across the Equator to spend their 'winter' (our summer) in the northern hemisphere, before returning south during our autumn in time to breed once again.

Chapter 5

Distribution and range

Habitats and their influence on behaviour

At first sight, a bird's habitat may not seem to have much direct influence on its behaviour, but a closer look reveals all sorts of subtle effects. For example, **wader** species that feed mainly on coastal marshes and estuaries are highly influenced by the twice-daily movements of the tide, as we have seen. The need to feed at low tide and roost at high tide influences their entire diurnal rhythms, which must be changed accordingly.

The feeding methods of **birds of prey** are also influenced by their habitat. In areas where they mainly prey on lightweight items such as **Ptarmigan**, **Golden Eagles** may nest fairly high up a mountainside; but in the west of Scotland, where they feed mainly on heavy prey such as Rabbits, they need to build their nest at low altitudes in order to carry their prey down the hillside rather than up, which would use more

▼ *Golden Eagles are one of the few species that can survive all year round on the high tops of Scottish mountain ranges*

▲ *Nightingales have a rich, complex song, which carries well through their dense woodland habitat*

energy. **Sparrowhawks** have evolved a compact shape with rounded wings and long tail in order to manoeuvre themselves through dense foliage; while falcons such as the **Hobby**, which hunts aerially, are more streamlined in shape.

Birdsong, too, is influenced by habitat. Indeed, it might even be argued that the different varieties of song have evolved to suit different habitats. Woodland species like the **Blackbird**, **Blackcap** and **Nightingale** tend to have rich, fruity songs, in order to penetrate the foliage around them to the greatest effect. With so many surfaces to absorb sound, a powerful song is the only way to ensure that the message is properly heard.

Species that live in marshes and reedbeds adopt a different strategy. Many of their songs, from species as diverse as **warblers** and **crakes**, are monotonous, rhythmic and repetitive. In fact they are often more similar to those of marsh-dwelling species of amphibians or insects than to other songbirds. Many years ago, the natural history sound recordist John Burton compared the songs of **Savi's** and **Grasshopper Warblers** with amphibians such as the Marsh Frog, and insects like the Wart-biter Bush Cricket, and concluded that their songs and calls had evolved in parallel, to suit the acoustic nature of the habitat: it seems that monotonous sounds are more effective in monotonous habitats.

◀ *The Grasshopper Warbler has a continuous, buzzing song, rather like an insect such as a bush-cricket*

▼ *The Mediterranean Gull has expanded its range northwards in recent decades to become a regular British breeding bird*

Finally, where birds choose to nest influences their behaviour and appearance to a large degree, especially regarding whether birds show sexual dimorphism (different plumages for males and females). For example, hole-nesting birds like the **tits** and **Tree Sparrow** can afford to have a brightly coloured plumage even in females, as they are not vulnerable to predators on the nest; while birds that nest in more open areas such as **House Sparrows**, **larks** and **pipits** tend to be brown and streaky in colour for effective camouflage. In open-nesting species where only the female incubates, such as **ducks**, the males are often bright and colourful while females are duller and cryptically marked; whereas large, aggressive species such as **Mute Swans** can afford to be showy and noticeable!

▲ *Like many ground-nesting birds, Skylarks have a brownish plumage which acts as camouflage against predators*

Range

The great ornithologist James Fisher once wrote that a bird does not have a range, only a 'current range'. He was referring to the fact that during as short a period as a single human generation – as little as two or three decades – the range of a particular species can alter dramatically, either expanding to colonise new areas, or contracting and disappearing from former haunts.

Even since the late 1960s, during the 40-plus years that I have been watching birds, I have witnessed a number of dramatic changes in the range and status of our breeding, migratory and wintering birds. Some of these, like the declines of farmland species such as the **Skylark**, **Yellowhammer** and **Grey Partridge**, are caused almost entirely by outside influences such as modern farming methods, and are therefore outside the scope of this book. Another, more positive change has been the welcome increase in the population and range of birds of prey such as the **Buzzard** and **Hobby**, this time due to the banning of harmful pesticides such as DDT, and legal protection against persecution. But other changes in range are more complex, and may be influenced either wholly or in part by the behaviour of a particular population of birds.

Perhaps the best-known examples of dramatic range changes are those sudden expansions of species such as the **Little Egret** and **Mediterranean Gull**, which have colonised Britain as breeding species in the last two or three decades. Again, human influence may be a factor, with climate change giving a helping hand to both these species. But other factors may also be at work: notably some kind of genetic mutation that allowed individual birds of each species to act as pioneers, followed by an eventual full-scale colonisation.

Another example involves a common British breeding species which now also winters here in good numbers. Twenty or 30 years ago wintering **Blackcaps** began to be reported in various parts of the country. At first most people assumed these were 'our' breeding birds

▲ *The Hobby, an agile, migratory falcon, has enjoyed a population boom in recent years*

◀ *Little Egrets are now a familiar sight in wetlands throughout southern Britain, having colonised from across the Channel in recent years*

▲ *Blackcaps are now regular winter visitors to Britain from Central Europe, thanks to a run of mild winters and the food provided by garden owners*

that had decided to stay put for the winter; but studies then revealed that these were actually German Blackcaps. Instead of migrating south-west to spend the winter in Spain, Portugal or North Africa, they had instead headed in a north-westerly direction and ended up in Britain.

Two factors enabled these birds to survive: first, mild winter weather prevented mass death by starvation; and second, the German birds adapted their feeding behaviour in order to take advantage of the plentiful seeds and other foods provided by us. As a result, they survived and thrived, and returned to Germany to breed ahead of their rivals. Twenty years later, and the entire population from this region now spends the winter in Britain.

As climate change adds yet another factor to the myriad influences on the range and distribution of Britain's birds, birders can look forward to many changes during the next few decades – some for the better, others for the worse. For example, **Bee-eaters**, which have already bred twice in the past decade or so, may become permanent colonists, while to the north we may lose Arctic species such as the **Snow Bunting**.

Chapter 6

Life and death

This may seem to be rather a forbidding section: the very title may put some readers off. And it is certainly true that not all the subjects covered lend themselves to careful field observation. Nevertheless, all these are part of a bird's lifecycle, and therefore vital to an overall understanding of bird behaviour in its broadest sense. I have tried to avoid too many technical terms (though some are inevitable), and where possible to give examples when a particular type of behaviour can be observed.

Moult

Birds do not keep all their feathers for the whole of their lifetime. Indeed, most undergo an annual moult, during which all or part of their plumage is replaced by spanking new feathers, giving the bird a better chance of survival.

Moult is vital for several reasons. The first, and most important, is to enable birds to fly as well as possible: old, broken or worn feathers reduce efficiency, and can lead to the bird being less able to find food, or more vulnerable to attack by a predator. Therefore having the best possible quality of plumage is absolutely vital. The second reason is that old, worn feathers are less efficient as insulators, and in cold winter weather the quality of a bird's plumage can make a major difference between life and death. Finally, old and worn feathers tend also to fade in colour and brightness, and for many birds, the quality of plumage, and in particular specific colours and markings, is what helps them to attract the best possible mate.

So for all these reasons, an annual moult is essential. But when should they carry out this potentially life-threatening process? The problem with even a partial moult is that during the intermediate stage when some feathers are old ones, and others new, the bird is at its least effective in terms of flight, keeping warm or attracting a mate.

For this reason, almost all birds undergo their major moult after the end of the breeding season, but before the onset of autumn and winter, when food resources are scarcer and the temperatures are lower. There are other advantages to moulting in late summer, especially for small **songbirds**: feeding a brood of hungry young will

▲ *These young Starlings have just fledged and left their nest; later they will moult into the familiar adult plumage*

have left their plumage in an exceedingly tatty state; there is plenty of foliage in which to hide to avoid predators; and there is a plentiful and freely available supply of food.

So don't be surprised when, after months of songbird activity in your garden or local park or woodland, everything suddenly goes quiet and the birds appear to have vanished. They are probably still around, just laying low!

Another group of birds that moults in mid to late summer are the **ducks**. Males of many species go into a state known as 'eclipse' plumage, a dull camouflaged body plumage superficially similar to the females, though usually with a few clues as to the bird's true identity. Male **Mallards** lose their glossy green head pattern, while male **Tufted Ducks** go from smart black-and-white to a dusky grey-brown and off-white colour scheme, rather similar to their mates. This often confuses inexperienced birders, who may think that all the males have vanished. The change occurs because during the moult the ducks become temporarily flightless, and so are vulnerable to predators – colourful, attention-grabbing plumage is a bad idea when you cannot fly. Then, from roughly early August onwards, the males reappear in their fine new feathers, as if by magic. Because of their vulnerability at this time, some ducks including the **Shelduck** gather in huge flocks for safety against predators.

Other birds are unable to enjoy the luxury of flightlessness, as they must constantly catch prey, so instead they moult gradually. Day-flying birds of prey such as **hawks** and **Common Buzzards** follow this pattern, gradually shedding and regrowing their flight feathers in systematic

▲ *Male Mallards adopt a plumage known as 'eclipse' when they moult during the summer months*

◀ *A juvenile bird such as this Black-headed gull can be tricky to identify, as it doesn't look much like its parents*

◀ *Juvenile Robins do not show the familiar red breast of the adult birds*

order, but never losing so many at once that their flight ability is compromised. For this reason some raptors may be in a virtually permanent state of moult.

Young songbirds hatched in the spring also undergo a period of moult: first from the downy covering into their first true plumage, known as 'juvenile'; then, usually a few weeks later, into the full adult garb. So baby **Robins** pass from downy fluff into the spotted browns and buffs of juvenile plumage, finally emerging, around two months after fledging, into the glorious adult plumage complete with orange-red breast.

For long-distance migrants such as the **Whitethroat**, moult is a vital prelude to achieving the best possible physical condition to travel thousands of miles south after breeding.

Some species undergo more than one moult a year, into and out of what used to be called 'summer' and 'winter' plumages, but are now usually referred to as 'breeding' and 'non-breeding'. Species that follow this pattern include **divers**, **grebes**, **gulls** and **terns**. As well as these two stages, gulls also go through a number of other plumages during the three or four years it takes them to reach full maturity. These moults can lead to confusion amongst birders as the differences between plumages can be quite subtle, and not all birds follow exactly the same stages.

Bathing, preening and feather care

Closely related to moult is the whole business of caring for feathers, involving a range of behaviours such as bathing (in dust and/or water), preening, and general feather maintenance.

Birds must bathe regularly in order to keep their plumage in tip-top condition, and free from dirt and parasites. They do so in a number of ways, of which the most commonly observed is simple bathing in shallow water, usually accompanied by a complex and choreographed set of movements designed to cover the whole of the bird's plumage with water in the most efficient and effective way possible. Virtually all species of bird bathe, but a small minority sometimes do so not in water, but in dust. At first this seems bizarre: after all, how can covering your feathers with dust keep them clean? Yet it does appear to work, apparently by removing oil, grease and parasites from the plumage. **House Sparrows** are particularly fond of 'dusting'.

Some birds, especially small songbirds, tend to bathe in a shallow puddle or the edge of a pond or stream; others, including **waders**, may do so in deeper water. True waterbirds such as **ducks**, **grebes** and **Coots** will bathe while swimming, submerging part of their body under the surface and letting the water wash over them until they are clean. Water is not always necessary to keep the plumage clean.

▼ *Birds such as this Chaffinch bathe regularly to keep their feathers clean and neat*

▲ *Cormorants need to hold their wings out to dry as, unlike ducks, they do not have waterproof feathers*

Some birds 'bathe' by allowing ants to swarm all over their feathers; the formic acid from the ants kills pests such as mites and ticks. Others use smoke, which has a similar effect.

After bathing, especially on a fine, warm day in spring or summer, birds often spend time sunbathing, enabling them to dry off their feathers and heat themselves up; **Blackbirds** are particularly fond of this. Other birds, notably **Cormorants**, hang out their wings to dry in the sun and wind; something they have to do because unlike those of ducks and other waterbirds, their feathers are not waterproof.

Following a bath, birds also spend time tidying up their plumage by preening, usually with their bills but occasionally also scratching with their feet. Preening is vital: it puts the feathers back where they should be, straightens out any problems, and allows the bird to remove any dirt or parasites which may have survived the bathing process. Some species, especially **waterbirds**, secrete oil from a gland near the upper base of the tail, which they use to waterproof and lubricate their plumage.

▲ *Barn Owls use their acute sense of hearing to hunt their prey in the long grass*

Sight, hearing and smell

Of all the senses, birds rely most on sight: a critical factor in the ability to find food, avoid predators, and assess a potential partner in courtship rituals. One of their most developed abilities is that of 'visual acuity' – or being able to differentiate between tiny differences – for example telling the difference between tiny particles of dirt and those of food. Even day-old chicks have been shown to be able to tell the difference between very similar objects, and to differentiate between subtle shades of colour.

Indeed, birds' colour vision is superb – another vital requirement, especially when searching for a mate, where tiny differences in the colour and brightness of a male's plumage enable the female to judge his state of health, and therefore whether he will make the best father for her young.

One way in which birds' eyesight differs greatly from ours is the position of the eyes. Most birds have eyes on either side of their face, which gives them excellent all-round vision for detecting food or predators, but means that they lack the binocular vision we take for granted, and which helps us judge perspective. The obvious exceptions to this rule are the **owls**, which have both eyes facing forward, enabling them to pinpoint and catch their prey more easily.

Another way in which birds differ from us is their ability to see ultraviolet light, enabling them to appreciate colours and shades that we are unable to see. This is especially useful for birds that feed on fruit and flowers, such as **hummingbirds**. The ability to detect ultraviolet light may also help migrating birds, especially when cloud cover obscures the sun.

Hearing is another vital sense for many birds, especially predators such as **owls**, which may hunt almost entirely using sound cues, especially if their prey is under snow or dense foliage. The ability to detect distant calls and songs is also vital, either to enable birds to hear warning calls as a predator approaches, or for females to hear distant males. Birds also have an extraordinary ability to differentiate sounds – especially useful for colonial-nesting seabirds such as **gulls**, **petrels** and **shearwaters**, where the returning adult finds the nest by listening for the young birds' calls.

Finally, smell is also used by particular groups of birds to find food. **Seabirds** have an especially well-developed sense of smell, a vital ability when food resources may be many miles apart at sea.

▼ Like other seabirds, Storm-petrels are able to find food at sea by smell

▲ *Manx Shearwaters spend much of their life out at sea, returning to their breeding colonies after dark, where they find their youngsters by listening to their calls*

Excretion

Just like any other living creature, birds must get rid of waste products, which would otherwise build up in their bodies and cause harm.

Unlike mammals, birds excrete their urine and faeces through the same opening, the cloaca, which is also involved in the reproductive process. Birds' excretory products vary considerably, depending on their diet: seed-eating birds produce dry droppings, those that eat moist food with a high water content, such as fruit and insects, produce wetter droppings. Sometimes droppings can actually be dangerous to our health, with colonies of **Feral Pigeons** in particular carrying lung diseases such as psittacosis, which in some cases can prove fatal.

Another way in which birds excrete is through their glands. True **seabirds** such as **shearwaters** and **petrels** have adapted ways of expelling excess salt from their food and the seawater they drink, which otherwise might build up in their bodies and kill them. They do so by means of glands just above their bills, which get rid of up to 90 per cent of the salt in their diet. **Owls** find it hard to digest the skin, bones and fur from their prey; so they get rid of these in the form of pellets which they cough up through their mouth.

Temperature regulation

Like all 'warm-blooded' animals, including humans, birds have the ability to regulate their body temperature by means of internal processes, rather than having to rely on external factors such as the sun, as do 'cold-blooded' creatures such as reptiles and amphibians. The process of temperature control is known as thermoregulation, and it enables them to manage their body temperature in response to changes in the outside environment.

Nevertheless, birds can suffer problems with both overheating and severe cold, and have adopted a range of different behavioural strategies to deal with these. Small birds such as **songbirds** are especially vulnerable to rapid changes in temperature, as their larger surface area to volume ratio means that their bodies lose or gain heat much more quickly than those of larger birds.

During spring or summer mornings, species such as the **Blackbird** will warm up their bodies by sunbathing, spreading out their feathers in order to gain maximum benefit from the sun's rays. Later in the day, if the weather gets very hot, they must reduce their body temperature or risk overheating. They may do so by bathing, but because they do not have sweat glands, they cannot sweat away moisture to keep cool as we do. Instead, they have to pant to allow heat to escape from

▼ *Blackbirds warm up on cool days in spring by sunbathing*

their throat and breathing passage. They may also seek shade, which explains why small birds are often very hard to see during hot summer days, staying in cover except at dawn or dusk when they emerge to feed.

Cold weather creates a very different problem: the rapid loss of heat from the unfeathered 'bare parts', such as the bill, legs and feet. To counter this many small birds will roost together, huddling up as close as possible in order to take advantage of their collective body warmth. During the short winter days birds retain heat by fluffing out their feathers, trapping pockets of warm air beneath. This may give them a very different appearance from normal, and make identification difficult.

Waterbirds, such as **ducks**, **geese** and **swans**, generally find temperature regulation easier, as during very hot weather they can simply immerse themselves in cooler water. In severe cold they conserve heat by standing on one leg on the ice, which reduces heat loss through their feet.

▶ *Mallards, like other waterbirds, may find it hard to find food when ponds and lakes ice over*

▲ *The Mistle Thrush is sometimes known by the old folk-name 'Stormcock', because of its habit of singing during bad weather*

Birds and weather

As the great American bird artist Roger Tory Peterson once said: 'Birds have wings – they travel.' By spending so much of their lives in the air, birds are surely influenced by the weather more than almost any other living creature.

As a result, birds have gained a reputation as excellent weather forecasters. They often alter their behaviour as a result of changes in weather conditions, so by observing this, our ancestors were able to predict the coming weather for themselves. Much of this knowledge has been passed down from generation to generation in rhymes and proverbs, which make up a unique body of weather folklore.

For example, the 'tumbling' behaviour of **Rooks** in autumn is supposed to foretell a change in the weather, probably because this behaviour tends to occur during windy conditions, which usually signify the coming of a depression. Insect-eating birds, such as **Swallows** and **martins**, also change their behaviour depending on the current weather: during settled spells of high pressure they feed on insects high in the sky, while during changeable periods of low pressure they tend to

▲ *Robins, like other garden birds, survive hard winters thanks to the food we provide in our gardens*

come lower, following their insect prey. By observing this behaviour it is possible to forecast the following day's weather with some accuracy.

Other species of bird are often associated with rain, such as **woodpeckers**, whose habit of calling and drumming before the arrival of bad weather has given them the name 'rain bird' in many parts of Britain, Europe and North America. In Shetland, the **Red-throated Diver's** habit of calling when rain is expected has earned it the folk name of 'rain-goose'.

Birds are also affected by extremes in weather, with harsh winter weather perhaps the most serious threat. Small birds such as the **Wren** and **Robin** are particularly at risk, as they must eat around one quarter of their body weight every single day if they are going to survive. When the ground is covered with a layer of snow, or when freezing temperatures cause branches and twigs to ice over, then they simply cannot get to the seeds or insect food they require. As a result, many species change their behaviour, with normally shy birds such as **woodpeckers** and **Nuthatch** becoming far more confiding, often visiting bird tables to get access to a ready supply of food.

During the breeding season the weather can also affect birds. For early breeders, such as the **Blackbird**, a late cold snap may reduce their food supplies at a crucial time. Later on in the spring, cool wet weather in May and June will reduce the chances of eggs hatching, and even if they do, the parents may not be able to obtain enough food to satisfy their hungry chicks. This is especially crucial for insect-eating species such as **tits** and **warblers**, or birds on the northern edge of their range in Britain, such as **Golden Oriole**.

▲ *Golden Orioles are one species widespread in continental Europe that may do well in Britain in the future because of climate change*

Every spring and autumn, migrating birds travel huge distances across the globe, as they seek out the very best places to breed and to spend the winter. Along the way, they encounter all kinds of weather, from helpful following winds to potentially fatal gales, storms and hurricanes. Many fail to survive the journey; those that do need a mixture of instinct, good luck and an ability to deal with weather systems, which has evolved over many generations.

In spring, returning migrants often delay their arrival by a week or more, as they wait for good weather over the Channel to allow them to make the crossing safely. Easterly winds may bring drift migrants across the North Sea, blown off course on their journey from mainland Europe to Scandinavia, to land on our east coast.

In autumn small birds run even greater risks from the weather. Migrants heading south from Scandinavia wait for a cold front, with clear skies and following winds, ideal for crossing the North Sea. But sometimes they get things wrong, encountering poor weather associated with depressions, and becoming disoriented. Many perish, falling exhausted beneath the waves. But others fly on through the wind and rain, making landfall on Britain's east coast, to the delight of birders dedicated enough to venture out in the bad weather. Even more extraordinary is the annual arrival of North American passerines in Britain, swept across the North Atlantic by strong westerly gales, to arrive exhausted in vagrancy hotspots such as the Isles of Scilly.

In the longer term, global climate change threatens to affect the lives of birds more than any other factor. We have already seen some species, such as **Great White Egret**, **Little Egret** and **Mediterranean**

▶ *Bee-eaters may eventually colonise Britain as a result of global climate change*

Gull, shift their breeding ranges northwards as a result of climate change. Breeding species such as the **Bearded Tit**, **Hobby** and **Nightjar**, once confined to southern Britain, are now beginning to extend their ranges northwards. Meanwhile, at the other end of the country, three birds of the high tops, the **Snow Bunting**, **Dotterel** and **Ptarmigan**, are likely to disappear as British breeding birds, as a result of major habitat change.

But every cloud has a silver lining, and in this case it comes in the form of new species colonising Britain from the south, and possibly also from the east. These may include such exotic creatures as the **Hoopoe** and **Bee-eater**, both of which now breed within reach of Calais, as well as less glamorous birds such as the **Great Reed Warbler**, and the **Black Kite**, one of the world's most adaptable and successful raptors.

If we do gain a more continental climate, then several species with a more easterly distribution, such as **Common Rosefinch** and **River Warbler**, could find eastern Britain a suitable place to colonise. Whatever happens, Britain's birders can look forward to an exciting time in the next half-century.

Bird feeders can spread disease amongst songbirds such as these Great and Blue Tits

Disease and death

Like all creatures, birds suffer their fair share of disease. Indeed, along with killing by predators, and lack of food, disease is one of the three major causes of death in wild birds. Those especially at risk include young birds just out of the nest, whose immune system may not be quite so well developed as that of the adults; and older birds, whose bodies may have been weakened by the toll of raising successive broods of young.

Another problem is epidemic diseases, which affect colonial species such as **seabirds**, and sociable ones such as **Starlings** and **House Sparrows**. Diseases such as salmonella have increased in recent years due to the artificial concentration of birds brought about by our enthusiasm for feeding them – a very good reason to keep your garden bird feeding station clean and replace old or mouldy food regularly.

Many bird species also carry unwelcome guests in the form of parasites. These may be either endoparasites such as liver flukes or tapeworms, which live inside the bird's body; or ectoparasites such as fleas, feather lice and mites that live on the outside, usually underneath the feathers. Many of these parasites are unique to

▶ *Like other seabirds, Gannets sometimes wash up dead on the shoreline, having perished at sea*

particular species of birds, and have co-evolved with them over many thousands of years. They may cause some harm, leading to death in a few cases, but for most host species appear to be accepted as an occupational hazard.

Occasionally an epidemic of disease will devastate a bird population, such as the waves of botulism that occasionally affect **waterbirds** of several different species, causing widespread death. Fortunately these outbreaks are relatively rare, and in the longer term most populations appear to bounce back from the epidemic, and recover their former numbers within a few years.

As to the difficult question of how long birds live, a general rule of thumb is that the larger the bird, the longer it is likely to live. A rough rule is that average longevity is correlated with body weight, using a complex mathematical formula. Using this, a species weighing approximately 32 times as much as another will live for around twice as long.

So most songbirds such as the **Robin** and **Blue Tit** have a mean longevity of only one or two years, and a maximum lifespan, in a very few cases, of perhaps seven to ten years. There are, of course, a few individual exceptions to the rule of live fast, die young: ringing recoveries have included a **Blackbird** and **Starling** aged 20 years, a 16-year-old **Swallow** and a 15-year-old **Great Tit**.

▲ *Oystercatchers are one of our longest-lived waders – one survived for at least 16 years*

Birds of prey such as **eagles** do not become fully mature for five or more years after hatching, and may live as long as 20 or even 30 years. Surprisingly, perhaps, the oldest recorded wild British birds are somewhat smaller in size: an **Oystercatcher** which survived for 36 years after being ringed (a good advert for a shellfish diet!), and a **Fulmar** which lived for an incredible 50 or more years.

Birds that generally live longer tend to follow a strategy of small clutch sizes followed by a long period of parental care (as in many **seabirds**), as opposed to large clutches and early fledging (as in most **songbirds**).

Finally, there are always exceptions to the longevity rule – for example when birds are kept in captivity. Safeguarded from predators, disease and other life-threatening factors, some can live as long as their owners. The record goes to a **Sulphur-crested Cockatoo** in London Zoo, known to have been more than 80 years old when it died in 1982.

PART TWO:

FAMILIES AND SPECIES

This section of the book deals with behaviour family by family, and species by species. There will inevitably be some overlap with Part 1, but by grouping together related species I am able to cover distinctive behaviour patterns such as flocking by tits in winter, or the different methods of feeding adopted by different groups of ducks. It covers both typical and some unusual aspects of behaviour, and includes the 200 or so species you are most likely to encounter in Britain.

Other reasons for grouping by family include:

- To provide an easy, quick reference to a particular group or species
- To give helpful hints and tips to aid identification
- To give a deeper insight into particular behaviours associated with particular groups of birds

Note: in some cases I have grouped similar families together for ease of use, such as the section on Seabirds, which includes unrelated species such as shearwaters and auks, which share the same habitat and are likely to be seen together. The same applies to the section on Swift, Swallow and martins.

DUCKS, GEESE AND SWANS

Ducks, geese and swans together make up a large and diverse group of birds generally known as wildfowl or waterfowl. These species have long been associated with humans – some are domesticated, while others are frequently hunted. Hence many species have developed an understandable wariness of human beings!

Ducks

Around 20 species of **ducks** are regularly seen in Britain, and can be put in different categories depending on their behaviour and habits.

First are the 'dabbling ducks', so-called because of their habit of feeding by working their bills along the surface, as well as ducking their heads under, or occasionally up-ending to take food. This category includes some of our most familiar species as well as one much less widespread one: **Mallard**, **Shoveler**, **Wigeon**, **Pintail**, **Gadwall**, **Teal**, and the scarcer **Garganey**.

Each feeds in slightly different ways: for example, the Shoveler sweeps its bill through the water, filtering out tiny morsels, whereas the Wigeon usually feeds on grass, by grazing on land. Gadwalls often accompany Coots, and appear to take advantage of the fact that by diving the Coot stirs up the water, bringing morsels of food to the surface for the Gadwall to feed on. Pintail, Teal and Garganey are generally shy, often flying when they detect your presence; whereas Mallards in town parks will generally allow a close approach – indeed most people's first experience of 'bird behaviour' is when they feed the ducks as a child! However, they may need to avert their eyes if the male Mallards display their frankly shocking mating behaviour, during which several males often gang up on a single female.

▲ *Pintail feed by 'up-ending' in shallow water to find aquatic plants and invertebrates*

▼ *Teal, like most waterfowl, struggle to find food when freezing weather makes ponds and lakes ice over*

▲ *Tufted Duck are 'diving ducks', which go below the surface of the water to find food*

Then there are the 'diving ducks': **Tufted Duck**, **Pochard** and **Scaup**. The first two species are common in Britain, especially during the winter when the breeding population is far outnumbered by immigrants from the north and east. Both usually live on inland waters, and dive for food, often going quite deep underwater. The Scaup is a more sea-going species, though can also be found on reservoirs and gravel pits, especially those near the coast.

Another group of diving ducks are sometimes referred to as 'seaducks', and include **Goldeneye**, **Eider**, **Common** and **Velvet Scoters**, and **Long-tailed Duck**. These, as their name suggests, are generally found around the coasts, where they often form large flocks offshore, sometimes consisting of several different species, all diving for food (mainly molluscs on the sea bed). Goldeneyes and Eiders can also be seen courting in early spring, throwing their heads back in display and uttering extraordinary calls.

Three other species are named after their characteristic bills with serrated edges: the 'sawbills'. The group includes **Red-breasted Mergansers**, **Goosanders** and **Smew**. Like the seaducks, they dive for their food – fish which are gripped in the saw teeth of the bill. Red-breasted Merganser and Goosander both breed in Britain, generally on fast-flowing upland rivers or lakes, though in winter Goosanders visit deep lakes and Red-breasted Mergansers are usually found near the coast or offshore. Smew is a winter visitor, and prefers gravel pits, though it is a shy bird and often vulnerable to disturbance.

Finally, there are three 'miscellaneous species': the **Mandarin Duck**, **Ruddy Duck** and **Shelduck**. Mandarin and Ruddy Ducks are both introduced species, and each displays some fascinating behaviour, especially during their courtship displays. However, the Ruddy Duck is now on its way out as a British bird, having been culled to prevent it flying off to Spain and interbreeding with the much rarer White-headed Duck. The Shelduck is not really a duck at all, but intermediate between ducks and geese. It is generally found around the coasts, especially on marshes and estuaries, where it may gather in large flocks.

▲ *With his smart black-and-white plumage, this male Eider is very different from the browner females*

▶ *Red-breasted Merganser is one of three British species of 'sawbill' ducks, so-called because of their distinctive slender bill with serrated edges*

▼ *Originally introduced here from China and Japan, the male Mandarin is one of the most colourful and distinctive of all Britain's birds*

Geese

Wild geese are among the best loved of all British birds, especially when they gather in huge flocks in winter. However, the introduced Canada Goose is one of the most loathed of all birds.

Eight species of goose live in or regularly visit Britain, of which four are collectively known as 'grey geese'. This category includes **Greylag**, **White-fronted**, **Pink-footed** and both 'tundra' and 'taiga' subspecies of the **Bean Geese**. The latter four species are all winter visitors to Britain, arriving in autumn from their breeding grounds to the north and east, and spending the winter months feeding in large, noisy flocks, generally on farmland but also on coastal marshes. The Greylag Goose breeds in northern Britain, but in recent years a large feral population has established itself in the south. Geese tend to be creatures of habit, feeding in their favourite areas by day, and as night approaches moving away to roost in large flocks nearby, often near water. Geese usually migrate in large flocks, flying in V formation (which increases efficiency and reduces exhaustion) with a leader in front, to guide them along the route.

Two other species of geese, **Barnacle** and **Brent**, are smaller and more distinctive in appearance. Barnacle Geese have similar habits to the grey geese, but as with Greylags a feral, non-migratory population is now at large in southern Britain. Brent Geese visit in winter and prefer coastal marshes and estuaries, where flocks feed on eelgrass and other vegetation. They can also often be seen on grassy areas such as golf courses and playing fields, especially at high tide.

The two remaining species are both introduced: **Canada** and **Egyptian Geese**. Both were originally brought here as ornamental wildfowl in the grounds of stately homes, but have since escaped and spread, and are both now firmly established as British breeding species. Canada Geese form large, noisy flocks, and can sometimes drive away

▼ *White-fronted Geese travel to Britain in autumn from their breeding areas to the north, and spend the winter here*

▲ *The Canada Goose is a familiar non-native species of waterfowl throughout much of lowland Britain*

other species of wildfowl. Egyptian Geese are spreading rapidly from their original release site in Norfolk, and are now at large over much of southern and eastern England. Despite their name, Egyptian Geese are not really geese at all, but are in their own separate genus. They breed early in the year (a hangover from their native home in Africa) and nest in holes in trees, from which the chicks jump soon after hatching.

Swans

Three species of swan are found in Britain: one resident, the **Mute Swan**, and two primarily winter visitors, **Bewick's** and **Whooper Swans** (though Whooper does breed occasionally in northern Scotland).

Mute Swans are one of our most familiar birds, and also our largest and heaviest bird. Although they often allow quite close approach, they can be dangerous, particularly to small children, though the oft-quoted remark that 'they can break a man's arm' is an urban myth. Mute Swans generally pair for life, and breed at the same site each year, building a huge nest out of twigs. Sadly these nests are often accessible to human vandals and animal predators, though if the pair is successful they usually manage to raise a large brood of cygnets. Despite their name, Mute Swans do utter various sounds; the name derives from the fact that unlike Whooper and Bewick's swans they do not usually call in flight.

Whooper and Bewick's Swans arrive in large flocks each autumn: Whoopers migrate from Iceland at incredibly high altitudes, while Bewick's make an even longer journey here from Siberia. Both enjoy the benefits of Britain's mild winter climate, where plenty of food is available. They usually gather in traditional sites, though small flocks can be found feeding in fields away from these. Dusk is a good time to see them in large numbers, as they gather together to roost for the night. At various Wildfowl and Wetlands Trust centres, such as Welney in Norfolk and Slimbridge in Gloucestershire, you can also see them being fed under floodlights during the winter months.

▼ *The Mute Swan is the largest and heaviest British breeding bird*

GAMEBIRDS

Ten species of gamebird breed in Britain, five of which were introduced here, a statistic which reflects these birds' importance as objects of quarry or ornament. Two ornamental species are Golden and Lady Amherst's Pheasants, both of which are highly localised – indeed Lady Amherst's may now be extinct as a British breeding bird, with just one recent sighting at its former stronghold in Bedfordshire.

Their better-known relative, the **Common Pheasant**, was also introduced, probably by the Romans, but for food rather than pleasure. It has since become our most widespread gamebird, and one of the commonest species of all, largely because up to 35 million young birds are released each year for shooting. Pheasants are birds of woodland and woodland edges, and are often very approachable, though when surprised they fly away noisily. Females are often more skulking, especially during the breeding season.

The two species of **partridge**, **Grey** (or English) and **Red-legged** (or French), are often seen together, though the introduced Red-legged is usually bolder and more inclined to sit out in the open. Both species can be very wary, and with good reason: partridge shooting is a popular sport. Early mornings and evenings are the best time to look for them, especially on the edges of fields near cover such as long grass and

▶ *The Lady Amherst's Pheasant, originally brought here from Asia, is now doomed to disappear as a British breeding bird in the next few years*

▲ *Grey Partridges are declining faster than almost any other British bird, due to loss of habitat and food on farmland*

hedgerows. Their tiny relative, the **Quail**, is a rare summer visitor to southern Britain, mostly in arable fields. It is hardly ever seen; if you come across it at all it is usually by hearing its distinctive call, sometimes transcribed as 'wet-my-lips'.

The four members of the grouse family are a contrasting bunch in appearance, habitat and behaviour. The **Red Grouse** is the species most often shot, and lives exclusively on open moors in northern and western Britain. Red Grouse are always very wary, often sitting tight before exploding in a flurry of wings and disappearing low over the horizon. Listen out for their calls if you want to locate them. **Black Grouse** and **Capercaillie** are both birds of the forest, though Black Grouse also live on more open moors.

Both species partake in 'lekking' in early spring, in which a number of males compete in a communal display in order to attract females. Sadly, because both species are now rapidly declining it is not advisable to visit their leks, as this can cause disturbance. Indeed, even walking through their habitat at this time of year is now actively discouraged. Your best chance to observe their extraordinary and wonderful behaviour is to visit a site where the birds can be viewed from a hide or the road.

The other grouse species, the **Ptarmigan**, is a bird of the high Scottish mountaintops. In winter it moults into a white plumage for camouflage amongst the snow, while in spring, summer and autumn it goes through phases of grey and brown to match the boulders amongst which it nests. Ptarmigans can be very approachable, but beware of disturbing them by approaching too closely.

DIVERS AND GREBES

Divers and grebes are both families of waterbirds, characterised by their ability to swim and dive. Unlike true wildfowl (ducks, geese and swans), grebes only have partially webbed feet. In both divers and grebes the legs are placed towards the rear of their body, making them well adapted to water but clumsy on land.

Two species of diver, **Red-throated** and **Black-throated**, breed in Britain, while the third, **Great Northern**, is a regular winter visitor. However, all three species are more likely to be encountered outside the breeding season, generally offshore, though occasionally on inland gravel pits or reservoirs. They are generally identified by their distinctive outline, with strong dagger-like bill, and a low profile in the water.

As their name suggests, they dive frequently, often for long periods of time, and are highly mobile underwater, sometimes reappearing a considerable distance from where they originally dived. Despite their appearance they are strong flyers, usually flying low over the waves in a very direct manner.

During the breeding season they nest at the edge of small lochs, where they may draw attention to their presence by their plaintive calls. Indeed, in Shetland, the Red-throated Diver is known as the 'rain goose' because of its supposed habit of calling when it is going to rain! Remember that nesting divers are easily disturbed, and never approach too closely.

▼ *The Red-throated Diver is the commonest member of its family to breed in Britain*

▲ *Our two smallest grebes: Little (top), and Black-necked (left, non-breeding; right, breeding plumage)*

Grebes are mainly birds of fresh water, especially during the breeding season. Two species, **Little** and **Great Crested**, are common British breeding birds, while the other three, **Slavonian**, **Black-necked** and **Red-necked**, are rarer and more localised. All grebes have ornate breeding plumage and impressive courtship displays, during which they 'dance' in the water, rising up almost vertically or pursuing each other in pair-bonding movements, sometimes offering each other pieces of weed as a token of their affection.

In winter all grebes apart from Little Grebe may also be found on the sea, usually close inshore. All five species dive for food throughout the year. During the breeding season Great Crested Grebes often cover up their eggs with weed to deter predators, a habit which leads to the pale eggs becoming stained a greenish colour. Once hatched the young will frequently hitch a ride on their parents' backs, always a memorable sight.

SEABIRDS (SHEARWATERS, PETRELS, GANNET, CORMORANTS, SKUAS, AUKS)

This collection of different families has one thing in common: they are generally found either around our coasts or out at sea (though in recent years the Cormorant in particular has ventured inland to breed).

Shearwaters and **petrels** are 'true' seabirds, spending virtually the whole of their lives at sea, and coming ashore only to breed. Four species breed in Britain and Ireland: **Manx Shearwater**, **European Storm-petrel**, **Leach's Storm-petrel** and **Fulmar**. The first three visit their island breeding colonies only at night, to avoid predators such as gulls, and as a result can be hard to see. All the petrels, like their relatives, are very long-lived; one Fulmar ringed on Orkney was known to be well over 50 years old.

Manx Shearwaters, as their name suggests, glide on stiff wings low over the sea, and can be observed doing so either from the coast or from a boat, especially during the evenings when they gather offshore before returning to their breeding colony. The two species of storm-petrel are more pelagic in their behaviour, and are only usually seen from land during onshore gales in autumn, when they may appear in quite large numbers. The European Storm-petrel is a tiny bird with a

▼ *Well over three-quarters of the global population of Manx Shearwaters breed in Britain*

▲ *The Gannet is Britain's largest seabird*

weak, fluttering flight; while its larger cousin Leach's Storm-petrel has a more deliberate flight action, often said to be like that of a nightjar, with flaps punctuated by long glides. The Fulmar superficially resembles a gull, but a closer look reveals the characteristic 'tube nose' which marks it out as a member of the petrel family, and in flight its superb aerobatic ability makes it a joy to watch.

The **Gannet** is Europe's largest and most impressive seabird, especially when seen at its huge nesting colonies off the Welsh, Scottish and northern English coasts. It is superbly adapted for aerial diving, plummeting from a great height into the water in order to catch fish. At breeding colonies you can also watch the birds displaying, squabbling and fighting over their tiny territories.

The **Cormorant** and **Shag** are coastal species, but the Cormorant is also a common sight inland, especially on rivers and gravel pits. Both species are accomplished divers for fish, but because they do not have waterproof feathers they must stand around afterwards drying their wings in a characteristic pose. Both are also communal breeders: Shags on cliffs with other seabirds, and Cormorants in a variety of places including trees, where they build untidy nests out of sticks.

Skuas are related to gulls and terns, but are more coastal in distribution than most of their relatives. Both British breeding species, **Great** and **Arctic Skuas**, feed by stealing from other birds, a habit known as kleptoparasitism. At seabird colonies they can often be seen chasing birds like Kittiwakes or terns in order to get them to regurgitate

▲ *Gannets nest in vast colonies numbering thousands of birds, known as gannetries*

and drop their food! If you venture close to a skua breeding colony watch out for another unpleasant habit: that of attacking intruders by flying towards them at great speed – a really frightening experience. Outside the breeding season skuas may also be seen offshore, from any of our coasts. Skuas are also fierce predators: Great Skuas will attack and kill smaller seabirds such as Guillemots and Kittiwakes.

Auks are the northern hemisphere equivalent of penguins, and have many similar characteristics – though unlike penguins they have not lost the ability to fly. Generally, though, they are rather ungainly in the air, and also unsuited to land; their real home is underwater, where they can dive to great depths in search of food. There are four British breeding species, **Guillemot**, **Razorbill**, **Puffin** and **Black Guillemot.** Guillemots and Razorbills nest on steep sea-cliffs, laying a single egg on a narrow ledge; Puffins make burrows, while Black Guillemots breed in holes or crevices amongst boulders. Outside the breeding season all auks become more pelagic in behaviour, heading out to the open sea. Occasionally a storm will drive them on to the coasts or even far inland, in what is known as a 'wreck'.

▲ *The Arctic Skua comes in two distinct colour forms: dark-morph (shown here) and light-morph*

▶ *The Puffin takes the prize as Britain's favourite seabird, even though many people have never seen one, as these birds nest on remote offshore islands and sea-cliffs*

HERONS AND EGRETS

The three regular British breeding species of heron and egret, Grey Heron, Bittern and Little Egret, are representatives of a large family of long-legged waterbirds found mainly in warmer parts of the world.

The **Grey Heron** is one of our largest flying birds, and is a familiar sight either overhead or standing stock-still by the side of a lake or river, hunting its aquatic prey. Herons fish mainly by stealth: waiting and watching, followed by a swift stab with their sharp, powerful bill. They nest in large colonies ('heronries'), and are very early breeders, often starting in January and having young in the nest by March.

The **Bittern** is one of our shyest and most elusive birds. Dwelling in reedbeds, it rarely shows itself, only occasionally appearing at the edge of the reeds, or seen briefly in flight as it travels the short distance from one part of its territory to another. In winter Bitterns can be more visible, especially during harsh weather, when they emerge from their reedbed home to feed. Another good time to look for Bitterns is during May and June when they are busy feeding young.

The **Little Egret** is a recent colonist as a British breeding bird, and like many of its relatives it nests colonially in trees. However, you are much more likely to see it hunting for fish and aquatic invertebrates at low tide in coastal harbours or areas of inland water near coasts. At high tide Little Egrets come together to roost, sometimes in quite large flocks.

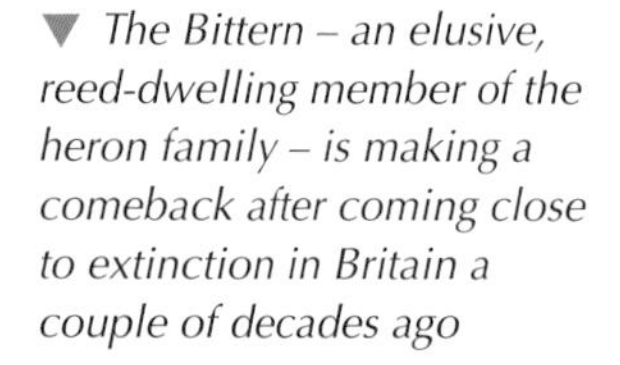

▼ *The Bittern – an elusive, reed-dwelling member of the heron family – is making a comeback after coming close to extinction in Britain a couple of decades ago*

▲ *Our commonest heron, the Grey Heron is found in a wide range of wetland habitats*

In recent years several other species of heron and egret, along with other large and long-legged wading birds, have bred here, and may be on the verge of colonising permanently. These include **Cattle Egret**, **Great White Egret** and **Little Bittern**, all of which have successfully bred on the Somerset Levels, and **Spoonbill**, **White Stork** and **Glossy Ibis**, which have bred or attempted to breed in eastern England. The **Common Crane** – Europe's tallest bird at well over a metre high – also now breeds regularly in East Anglia after an absence of more than 400 years, while released birds have now begun to breed in the West Country too. Soon, perhaps, we will witness the spectacular courtship dance performed by flocks of Cranes at their breeding sites.

RAPTORS

Raptors, or day-flying birds of prey (excluding owls), are a diverse group, which includes some of our best-known and most magnificent species, such as eagles, hawks and falcons. Fifteen species breed in Britain, though a number of these have a very restricted range, and are unlikely to be seen without special effort.

Starting with the largest raptors, two species of **eagles** breed here: **Golden** and **White-tailed**. Golden Eagles are mainly confined to Scotland and Ireland, though a lone male still holds territory in the Lake District. These are magnificent birds, which cover a vast territory and are never easy to find. In a suitable area look out for them soaring on huge wings over crags and mountains, though bear in mind that they tend to nest at lower levels than you might expect, to avoid having to carry prey upwards! Like all large birds of prey they make use of thermals and are more likely to be seen soaring in warmer weather. Golden Eagles usually have two young, but usually the larger one ends up getting the majority of the food, so that its smaller and weaker sibling starves to death, in a process known as 'Cainism' after the Biblical story of Cain and Abel, in which one brother slew another.

The White-tailed or Sea Eagle went extinct as a British breeding bird in the early 20th century, due to persecution. Since being reintroduced, the species is now thriving on the rocky shores of western Scotland and the outlying islands, and is now being reintroduced to areas on the east coast of Scotland as well. White-tailed Eagles regularly follow fishing trawlers, consorting with the much smaller gulls to take a share

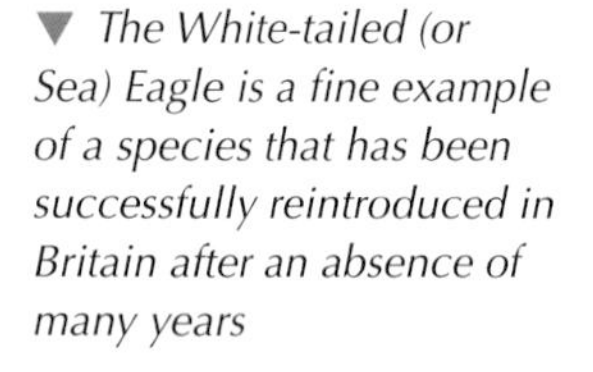

▼ *The White-tailed (or Sea) Eagle is a fine example of a species that has been successfully reintroduced in Britain after an absence of many years*

▲ *The Osprey (left) feeds on fish, while the Common Buzzard (right) has a wide-ranging diet including rabbits, earthworms and carrion*

of the fish thrown overboard, and can also be seen loafing around on mudflats, especially on the Hebridean island of Mull, where there are watchpoints to view the nests too.

The **Osprey** is our other fish-eating raptor, catching its prey by a feet-first plunge over lakes or shallow seas. As long-distance migrants, Ospreys are regularly seen in southern Britain on their journeys to and from Africa, sometimes hanging around wetlands for days or even weeks. They often perch on prominent posts or trees. Most breed in Scotland but there are a few pairs in England and Wales.

The **Common Buzzard** is our commonest large raptor, and can often be seen on a fine day soaring high over woods, especially in western Britain. Sometimes mistaken for an eagle, its wing-shape is in fact quite distinctive: it usually holds its wings bowed with the tips upward, and occasionally even hovers rather clumsily. The similar-looking but not closely related **Honey-buzzard** is a rare and shy bird that feeds mainly on the grubs of wasps and bees. It is a summer visitor; the best chance of seeing one is to visit known breeding sites in late May when the birds are displaying. Unlike the Common Buzzard it generally soars on flat wings.

Harriers and **kites** are superficially similar birds, with long wings and a habit of flying low over the ground. The **Red Kite** gave its name to the child's toy, and you can see why: this is one of our most graceful and acrobatic birds of prey, often seizing food from the ground without landing. Like Common Buzzards they will sometimes hover as they

▲ *The female Montagu's Harrier looks very different from her grey-coloured mate*

look for their prey. Having been reintroduced into parts of England and Scotland, Red Kites are now thriving and can even be seen coming into gardens for food in areas such as the Chilterns.

Marsh Harriers are birds of reedbeds and other wetlands, and like all harriers they fly low on V-shaped wings while hunting. **Hen Harriers** breed on moorland, though winter in wetland areas. Both species are often best seen coming into reedbed roosts in late afternoon during the winter. **Montagu's Harrier** is a very rare British bird, generally nesting in open farmland, over which it hunts on narrow wings that give it a buoyant flight.

The two **hawks** are both birds of wooded and forested areas. The commoner **Sparrowhawk** has made a comeback since declining due to agricultural pesticides that accumulated in its prey, and is a familiar sight in towns and suburbs. Often merely glimpsed as it passes by, its fast, low flight, using hedges or buildings as cover, is designed to surprise small birds. Sparrowhawks also have short, rounded wings and a long tail to manoeuvre through foliage, which gives them a very distinctive flight action out in the open: a short series of flaps followed by a glide. If you are lucky, a Sparrowhawk may visit your garden and seize an unsuspecting songbird before your very eyes, then sit nearby and pluck it ready for eating. **Goshawks**, by contrast, are generally elusive and hard to see, despite their large size. They are forest-dwellers, and the best way to catch sight of them is in late winter and early spring, when on a fine day pairs will display above the forest canopy.

▲ *Once confined to a small area of mid-Wales, the Red Kite has now successfully been reintroduced to much of lowland Britain*

The **falcons** are distinct from other raptors and belong to a different taxonomic order. Four species breed in Britain, of which by far the most common (though now sadly declining) is the **Kestrel**. Kestrels are generally seen hovering, a high-energy hunting strategy that enables them to catch their favourite prey of voles. They may also be seen soaring, gliding or simply sitting on a post or in a tree resting or watching out for prey. Our largest falcon, the **Peregrine**, hunts even more spectacularly: flying high in the sky before plummeting down at great speed onto an unsuspecting pigeon or other bird. Peregrines are the fastest moving creatures in existence, reaching speeds of more than 200 miles per hour when 'stooping'. They can also be seen hunting low over marshes in winter and moors in summer, often putting other birds into a blind panic. In recent years they have moved into city centres and now nest on large buildings such as church spires, from where they sit and survey their territory.

Our smallest falcon, the **Merlin**, nests on moorland and winters on the coast. It is another opportunistic hunter, which like the Peregrine chases and catches its prey with an amazing turn of speed. Finally, the **Hobby**, a summer visitor to Britain, hunts either by hawking for dragonflies (which it grabs in its talons and transfers to its bill to eat, hardly breaking its flight pattern) or searching for flocks of Swallows and martins, which it chases and seizes in flight. Look out for Hobbies on fine days in late April and May when groups often come together to hunt, especially over wetland areas.

RAILS AND CRAKES

There are just five species of rails and crakes found in Britain, showing very different behaviour. Two species, Coot and Moorhen, are relatively open in their habits; while the remaining three, Water Rail, Corncrake and Spotted Crake, are shy and elusive, with the latter two species often proving almost impossible to see.

▲ *Coot (left) and Moorhen (right) are both members of the rail family that have adapted to life on the water*

The **Coot** is such a common and widespread species in Britain it would be easy to take it for granted. Yet when watched closely its behaviour is fascinating, especially during courtship, when males will fight sometimes to the death. They do so by leaning back into the water and using their feet and claws as weapons. During the breeding season the Coot is one of the easiest birds to observe without disturbance: its nest is easily found, and once the eggs have hatched the chicks and adults appear to be well used to human beings. The young often beg their parents for morsels of food, and the pair can have their work cut out keeping the hungry chicks satisfied. On land or in flight Coots appear relatively clumsy.

The **Moorhen** shows very similar behaviour to the Coot, though instead of diving for food it picks items from off or just beneath the water's surface. Moorhens also frequently feed on land, often on areas

▼ *The Spotted Crake is one of the most elusive of all British birds, but can sometimes be seen well on migration in autumn*

of damp grass by water. As they walk they often bob their tail and head in a characteristic manner. Moorhens rarely seem to fly, preferring to paddle rapidly across the water or dive into cover when alarmed. Oddly, Moorhens are sometimes seen climbing trees, and on occasions even nest there. During the breeding process, the youngsters from an earlier brood sometimes help their parents with the new family.

The **Water Rail** is a much more terrestrial bird than its two aquatic cousins, generally hiding away in dense reedbeds or other fringe vegetation around the edge of ponds or lakes. Its long legs and large feet do allow it to wade and swim, but it is really designed to squeeze through narrow gaps in reeds, its body being laterally compressed to enable it to do so quickly and easily. Water Rails hunt their prey avidly, spearing it or seizing it in that long, sharp bill. They rarely fly, generally flapping a few yards before reaching cover. They do not appear to fear humans and can give excellent views, especially if you are patient and prepared to wait.

The **Spotted Crake** looks much like a smaller, stouter-billed Water Rail, and inhabits the same semi-aquatic habitat of dense vegetation. However, it is even more skulking, and often the only clue to its presence is its characteristic whiplash call. It is a long-distance migrant, wintering in tropical Africa.

The **Corncrake** was once known as the Landrail, which gives a clue as to its terrestrial lifestyle. Once widespread throughout Britain, it has suffered from the spread of modern farming and is now confined to the extreme north and west of Scotland and parts of Ireland. There, its call may be a ubiquitous 'sound of summer', but seeing the bird presents a far greater challenge. Corncrakes are ventriloquial, so pinpointing their position in long vegetation is almost impossible; best to wait patiently and hope the bird eventually shows itself. Early in the season, when the birds arrive back from Africa and the vegetation has not yet fully grown, is the best time to try to get a good view.

▼ *The Corncrake was once found across much of Britain, but is now largely confined to the extreme north-west of Scotland*

WADERS

There are 30 or so species of waders that regularly breed, migrate through or winter in Britain. Dividing them up into groups on behavioural factors is bound to be somewhat artificial, but for ease I have chosen a range of categories, some artificial, others based on classification.

Plovers

Plovers are a diverse group of waders, which nevertheless share some obvious characteristics, including a short, straight bill, usually used for picking items of food off the surface of mud or the ground; shortish legs and a characteristic 'stop-and-start' running action; and long wings, often used for epic migratory journeys.

The two smallest plovers in Britain are the **Ringed** and **Little Ringed**, and they have very similar feeding habits. The Ringed Plover tends, however, to be a bird of coastal areas, often found in the company of other small waders, though preferring to feed singly. The Little Ringed Plover colonised Britain between the wars, using newly dug reservoirs and gravel pits as nesting areas. Both species have very interesting breeding habits, including the famous 'distraction display', in which the bird will call plaintively while dragging its wing along the ground to draw predators (or human beings) away from its nest or chicks.

Two larger species, **Grey** and **Golden Plovers**, also form a 'pair'; however, despite their similarity of appearance they have rather different habits. Grey Plovers are winter visitors to coastal marshes, usually seen singly or in loose groups; whereas outside the breeding season Golden Plovers form flocks, often with Lapwings. During the breeding season Golden Plovers will also perform the distraction display to ward off predators on their moorland habitat.

▶ *The Ringed Plover has a short, stubby bill which it uses to pick up morsels of food from the surface of mud*

▲ *With its pied plumage and distinctive crest, the Lapwing is one of our most easily identified breeding birds*

The **Lapwing** is many people's favourite wader, and with good reason: as well as being stunningly beautiful, it also shows fascinating behaviour. Outside the breeding season Lapwings form huge flocks, gathering on farmland or coastal marshes to feed. They do so in characteristic plover style: taking a few rapid steps forward, pecking briefly at a morsel of food, then running forward once again, constantly on the lookout for more to eat. When disturbed Lapwing flocks rise up into the sky calling in alarm. When courting, Lapwings perform a wonderful aerial display, tumbling through the air like acrobats. Their young are, like all wading birds, precocial (able to leave the nest immediately) and cryptically coloured to avoid being spotted and caught by predators.

The final representative of the plover family is the **Dotterel**. Confined as a breeding bird to the highlands of Scotland, the Dotterel is occasionally seen in southern Britain as a migrant, on its way back in spring from its African winter quarters. Amazingly flocks (or 'trips')

▼ *The Dotterel is unusual in that the female has a brighter plumage and takes the lead in courtship*

of Dotterels are creatures of habit, often turning up in the very same field, on virtually the same day, from spring to spring. Once they reach their breeding grounds Dotterels live up the origin of their name (it means 'stupid fool') by being extraordinarily tame, often approachable to within a few feet (though this is not advisable when the birds are breeding).

Small waders

This is a diverse group whose species have little in common, behaviourwise, apart from their size.

The smallest of all are two species of **stint**: **Little** and **Temminck's**. These tiny waders are most often encountered on migration, as they feed frantically to build up fuel before the next leg of their journey between the Arctic and Africa. Another long-distance migrant, **Curlew Sandpiper**, whose name derives from its decurved bill, may accompany them. Like its much commoner relatives, **Dunlin**, **Knot** and **Sanderling**, it is a consummate traveller. Dunlins and Knots gather in huge flocks to feed and roost, crowding together for safety against predators as the waters rise at high tide. In flight, flocks of Knots appear to be controlled by some unseen hand, as they twist and turn through the air with extraordinary coordination. Sanderlings gather in much smaller groups on the tideline, racing away from the incoming water like little clockwork toys, their legs going like the clappers!

▼ *Our smallest regularly occurring wader, the Little Stint is a passage migrant through Britain, mainly seen in late summer and early autumn*

▲ *Waders such as Knots gather in vast flocks, especially when high tide forces them off their feeding areas and towards a roost*

▼ *Common Sandpipers are often found on streams and rivers, where they bob up and down while searching for food*

Two distantly related species, **Purple Sandpiper** and **Turnstone**, are often seen together on rocky shores, mainly in winter. They use their short, powerful bills to pluck molluscs and other invertebrate prey from rocks. Turnstones are often seen in coastal resorts, where they can be amazingly tame, even scrounging chips from bemused holidaymakers.

Three other species, **Common**, **Wood** and **Green Sandpipers**, are usually associated with fresh water, though on migration they also occur near the coast. Common Sandpiper invariably bobs up and down when feeding, a useful identification point (though Green can do the same thing). Green Sandpipers tend to sit tight until flushed, and then fly away calling noisily; whereas Wood Sandpipers behave more like a small 'shank', wading in deeper water with an elegant movement and action.

Medium waders

These four medium-sized waders are often encountered in a range of habitats and locations. The three 'shanks', **Redshank**, **Greenshank** and **Spotted Redshank**, are all long-legged waders with fairly long bills, and usually feed on mud or near the edge of water. The Redshank is the classic 'all-purpose' wader, adapted to a range of habitats, especially outside the breeding season. Its habit of taking flight and calling in alarm as soon as it is approached has earned it the nickname 'sentinel of the marsh'. In the breeding season Redshanks call incessantly, often perching on fence posts to get a better view of danger. Greenshank and Spotted Redshank are both more specialised feeders, with the Spotted

▼ *Ruffs gather during the breeding season at a 'lek', but this is a rare sight in Britain as very few pairs nest here*

Redshank's long legs and bill enabling it to wade quite deep into the water, making it appear more like a godwit in habits.

The 'odd man out' is the **Ruff**, which can bear a superficial similarity in structure, build and habits to the Redshank, especially outside the breeding season, when it shares the same habitat. Ruffs also feed in drier areas, however, such as ploughed fields. In the breeding season the Ruff's behaviour is radically different: males (which are much larger than females) develop a splendid headdress and gather in leks to try to woo the females. After mating with her chosen male, each female incubates her eggs and rears the chicks alone.

Large waders

This collection of birds includes the two **godwits** and the **Curlew** and **Whimbrel**, as well as oddities such as the **Stone-curlew**, **Oystercatcher** and **Avocet**.

The two species of **godwit**, **Bar-tailed** and **Black-tailed**, are both large, long-legged wading birds characteristic of coastal wetlands. They generally gather in flocks, Black-tailed feeding methodically in deep water, while Bar-tailed prefers sandy shores and mudflats.

The **Curlew** and **Whimbrel** both have long, decurved bills that they use to probe into mud or soil to find invertebrate food. During the breeding season both have delightful display flights, uttering their haunting calls as they fly overhead. In winter Curlews may gather in quite large flocks on estuaries or mudflats, or to feed in flooded fields. Whimbrels migrate to Africa, often stopping off at coastal sites on their

▼ With its long legs and bill, the Bar-tailed Godwit is able to feed in shallow water and probe deep into the mud for food

way to refuel. In spring they may also be seen in fields as they stop to feed on their journey north.

The **Stone-curlew** is not a curlew at all, but a member of an African family known as the 'thick-knees' due to their peculiar anatomy. It is a bird of dry farmland and heath, and its large, staring eyes give away that it is mainly a crepuscular feeder. Stone-curlews are brilliantly camouflaged, especially during the day when they crouch in furrows, and are often only visible when they run from spot to spot on their long legs.

The **Oystercatcher** is a classic bird of coastal areas, found in a variety of sandy and muddy habitats where it can form huge flocks, especially at high-tide roosts. In the breeding season it frequents very different habitats including the edge of lochs and rivers, or grassy fields, sometimes far inland. Their 'piping display', in which several birds will conduct what looks like a coordinated dance while uttering loud, piping calls, is a great spectacle.

Surely our most elegant wading bird, the **Avocet** has an extraordinary upcurved bill, which it uses in a unique way. Instead of poking, picking or probing the mud, it sweeps its bill from side to side through water to filter out tiny aquatic organisms. Avocets often feed in groups, but during the breeding season they can get very territorial and aggressive when an intruder or potential predator comes near.

Snipe and Woodcock

Snipe, **Jack Snipe** and **Woodcock** are three species with long bills, short legs, cryptic plumage and similar feeding habits.

The **Snipe** is generally found in damp grassy areas, wet meadows or marshes, feeding close to cover and probing its very long bill deep into the mud. It is easily alarmed, flying away fast and high on rapidly beating wings. During the breeding season it has an amazing 'drumming' display, in which it flies high into the sky before plummeting earthwards, making a strange noise by vibrating its tail feathers.

Its smaller relative, the **Jack Snipe**, is far less easily seen, usually sticking very close to cover. One way to tell the two species apart is the Jack Snipe's characteristic bobbing action, as if its body is mounted on springs. The Jack Snipe stays put almost until it is trodden on, before flying silently away very fast and low, and soon taking cover.

The **Woodcock** is, as its name suggests, a bird of woods and forests, where it can be very difficult to see. Birds generally sit tight, relying on their cryptic plumage for protection. Your best chance of observing a Woodcock is during the 'roding' display on a fine evening in spring or early summer, when males fly around the tops of trees flicking their wings and calling; or in autumn and winter, when feeding birds may be flushed from a damp patch in a wood.

▶ *Snipe are able to probe their long bills deep into the earth even when the ground is covered with snow*

▼ *This Woodcock is beautifully camouflaged as it crouches down amongst dead leaves on the forest floor*

GULLS AND TERNS

Seven species of gulls and five terns regularly breed in Britain, while a number of other species occur on passage or as regular winter visitors. There was a time when birders largely ignored gulls, perhaps because they are common and familiar, but also because they exhibit a bewildering range of different plumages that can be confusing. Yet gulls are one of these easiest groups of birds of all to study, and close attention really does pay off.

The species most likely to be encountered, especially inland, is **Black-headed**, which has adapted so well to living alongside humans it is now a familiar sight in a whole range of habitats including gardens. Other species such as **Common**, **Herring** and **Lesser Black-backed** are also increasingly found inland, with the two larger species even nesting on city roofs and feeding on landfill sites nearby. Watch out for all four species flying into their roosts, especially on winter evenings when the numbers involved can run into the thousands.

Nesting colonies of gulls are also fascinating to watch, with every kind of breeding behaviour including complex courtship displays, mating, territorial squabbles and of course the raising of chicks, an avian soap-opera free for anyone to enjoy. Larger gulls, including Herring and **Great Black-backed**, often hang around on the edge of Black-headed Gull colonies, ready to nip in and grab an egg or a chick that has been left alone for a moment or two. Watch also for parent birds returning to their young: they go through ritualised bonding before the adult finally regurgitates their catch for the hungry chick.

The **Kittiwake** is one of the most marine of all gulls, nesting mainly on rocky cliffs, though also sometimes choosing warehouses on docksides. It is a sociable bird, and at its large colonies it constantly utters the cry that gives it its name.

▼ *Kittiwakes nest on narrow ledges, yet still manage to do their courtship displays there*

▲ *The Great Black-backed Gull is the largest species of gull in Britain*

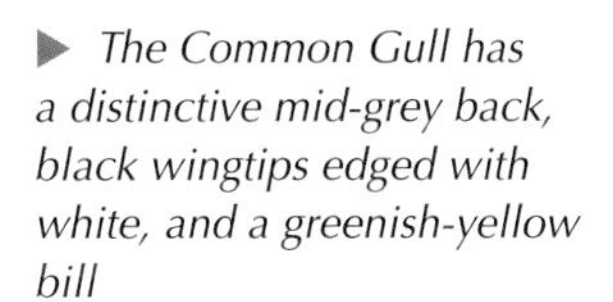

▶ *The Common Gull has a distinctive mid-grey back, black wingtips edged with white, and a greenish-yellow bill*

▼ *The Mediterranean Gull is a relatively recent colonist to Britain, having first bred here in the late 1960s*

▲ *The largest British breeding tern, this Sandwich Tern is bringing back food for its hungry chick*

Gulls can also be watched feeding, either opportunistically or when following a fishing boat to catch cast-offs. Looking closely through a feeding flock is a good way to pick up rarer species such as **Glaucous** and **Iceland Gulls**.

Two other species of gull are regular passage migrants or visitors: **Mediterranean** and **Little Gulls**. The former species, though scarcer, also breeds here in small but increasing numbers, usually among breeding Black-headed Gulls. Little Gulls have a very buoyant, rather tern-like flight.

Terns are, as one writer once said, 'gulls that have died and gone to heaven'. With their buoyant flight and graceful appearance they really do put clumsy gulls to shame. **Common Terns** are now often encountered inland, either on passage or breeding, especially on islands or artificial nesting rafts on gravel pits and reservoirs. Like gulls they have complex courtship rituals. Their close relative, **Arctic Tern**, has a much more coastal distribution and nests in large colonies, usually on offshore islands. If you make the mistake of getting too close the defensive parents will often mob you, even drawing blood with that dagger-like bill. **Sandwich Terns** are also birds of the coast, more gull-like in appearance than their relatives. **Little Terns** are delightful little birds, which nest on undisturbed shingle beaches around our coasts. Like other terns they can be watched hunting for food by plunge-diving into water. **Roseate Terns** are the rarest of our breeding species, found in a few colonies off the north-east coasts of England and Scotland.

PIGEONS AND DOVES

There really is no clear difference between pigeons and doves, apart from the fact that doves tend to be smaller and more delicate. There are five British species, three of which are almost ubiquitous: the Woodpigeon, Collared Dove, and the street Feral Pigeons which descend from domesticated Rock Doves. A few wild Rock Doves still live on remote coasts in northern and western Scotland. The Stock Dove is widespread but often overlooked, and the Turtle Dove, the only migratory member of its family in Britain, is much scarcer than formerly.

The **Feral Pigeon** is one of the commonest British birds, yet paradoxically it is virtually ignored by birders, thanks presumably to its dubious origin as a domesticated species. Yet if you want to observe bird behaviour at close quarters it is one of the best species to choose: widespread, used to humans and fascinating to watch. In early spring look out for males performing their courtship display to nonchalant females. To see the wild ancestor of this bird, the **Rock Dove**, you will need to travel to outlying islands and headlands in north-west Scotland, where the last remaining purebred birds still hold their own. These are much shyer than their feral descendants, flying away as soon as you get near.

The **Woodpigeon**, originally a bird of woodlands and farmland, has adapted very well to living alongside humans in cities, towns and gardens. It is not the most graceful of birds, but like all successful species has learned to be versatile in its feeding and breeding habits.

▼ *The Wood Pigeon is one of the commonest birds of the British countryside*

▲ *The delicate and beautifully-marked Turtle Dove is now threatened with extinction as a British breeding bird, thanks to modern farming methods*

A relatively recent colonist from Europe (having arrived in the 1950s), **Collared Doves** have become a familiar bird of towns and suburbs, especially well-wooded ones. They often visit bird tables, though like all pigeons they remain constantly wary, always on the lookout for danger. They nest throughout the year, and can have eggs or chicks in the nest in the middle of winter.

Stock Doves may prove to be hard birds to find; they keep themselves to themselves for much of the year, though on fine days in spring look out for pairs displaying over woodland. They will also visit gardens in some areas. Outside the breeding season Woodpigeons, Feral Pigeons and Stock Doves may join together in flocks to feed, especially on farmland.

The smallest British member of the family, the **Turtle Dove**, is a very shy bird, best detected by its purring call in May and June, after it has arrived back from Africa. Turtle Doves may be seen more easily on autumn migration, when they leave woods and heaths and feed on farmland and coastal areas. However, in the past few decades the population has plummeted and the Turtle Dove is now absent from many of its former haunts.

OWLS, CUCKOO AND NIGHTJAR

This miscellaneous group of non-passerine birds includes our largest family of nocturnal birds, the owls, with five British breeding species; and the only British representatives of two worldwide families: the nocturnal, summer-visiting Nightjar, and another summer visitor that lays its eggs in other birds' nests, the Cuckoo.

Owls are difficult to observe, largely because they are either nocturnal or crepuscular in habits, and even those species that do fly by day may be hard to find. Once seen, they are unmistakable: a combination of shape, forward-facing eyes and behaviour marks them out as unique. Owls have adapted to fit specific habitats, and as a result it is rare to see more than one species in the same place. During the daytime, a roosting owl can often be located by watching and listening for small birds agitatedly mobbing the intruder in their midst.

Our commonest species, the **Tawny Owl**, is also one of our most secretive: it is highly sedentary, spending most of its life in the same small territory, which it gets to know very well indeed. This is essential for a bird that is almost entirely nocturnal. As a result the best way to find this species is to listen for its characteristic hooting and ke-vick calls in late winter or early spring, when birds are marking out their territory prior to breeding. Once you have found a territory look for suitable nesting holes, usually halfway up a mature tree, and hope that you are lucky. Another way to find Tawny Owls, especially outside the breeding season, is to search for a roosting bird. During the daytime Tawny Owls sit tight, usually in a hollow in a tree trunk, and once you have found them may be easily visible.

▼ The Tawny Owl is the commonest British owl, and also one of the most sedentary and nocturnal

Barn Owls exploit a quite different habitat: open farmland, ideally of a more traditional nature, with older buildings where they can enter and make their nest. They are often seen at dawn or dusk, hunting like white ghosts over fields and marshes in search of their favourite prey of voles. With their soft plumage designed to allow silent flight, Barn Owls are especially vulnerable to getting wet and may miss a night's hunting if there is heavy rain. So if it has rained for a day or two and then stopped, that is a good time to look out for them as they will be eager to resume hunting. If you do find one you will marvel at its silent, slow and agile flight, with a short, strongly flapping hover preceding the plummet to earth to catch its prey.

Little Owls are the most diurnal owl species (along with Short-eared). They were introduced to southern Britain from continental Europe during the nineteenth century, and have become an established and generally welcome member of our avifauna. Little Owls frequently perch on the sides of trees (especially oak), or on stumps, fence posts or the roofs of farm buildings, where they sit and wait before plunging down to catch their prey of insects, worms and small rodents. They are often found in parks, though an early morning visit is a good idea as too

▲ *The Barn Owl hunts by floating low over the ground, on soft, silent wings*

▼ *The 'ears' on a Long-eared Owl are in fact tufts of feathers, and do not help the bird hear*

much human activity around may disturb them, causing them to retreat to their hiding places.

Long-eared and **Short-eared Owls** form a 'species pair', yet have quite different habits. Like the Little Owl, Short-eared is primarily diurnal, and will hunt low over moors and marshes on long, lazy wings, often looking more like a harrier than an owl. This makes it probably the easiest owl species to see well. Long-eared Owls are a complete contrast: almost entirely nocturnal, and very hard to see. However, you may be lucky enough to discover (or find out about) a daytime roost, where up to a dozen birds will spend the daylight hours huddled together in dense scrub.

The **Nightjar** is another difficult bird to see: you have to find a suitable lightly wooded heath or young conifer plantation, mainly in the south of Britain, and wait patiently at dusk – preferably on a fine evening between mid-May and July. If you are lucky, you will witness the incredible sight of this extraordinary bird gracefully hawking for insects, or a male in display, flashing the white patches on his wings while uttering his churring song. During the day Nightjars roost on heather or on the ground, and should not be disturbed.

The **Cuckoo** is our only parasitic breeding bird, each female laying up to 20 eggs in different nests of her host species, a strategy which maximises the chances of breeding success. Cuckoos parasitise a particular host species (the one in whose nest they were born), the three most common in Britain being Meadow Pipit, Dunnock and Reed Warbler. The female Cuckoo ejects one of the host's eggs before depositing her own.

▲ *Nightjars hunt by night, using their exceptional eyesight to catch moths and other insects*

◀ *The Cuckoo is the only British bird that lays its eggs in other birds' nests*

Once hatched, the Cuckoo chick grows very rapidly, ejecting any remaining eggs or chicks, and being fed frantically by the unsuspecting host parents until it becomes far too big for its tiny nest. Meanwhile the parent Cuckoos depart for Africa in June or July without ever seeing their offspring, which having fledged manages to find its way south by itself. Watch out for Cuckoos in late April or early May, when males have just arrived back and are easier to see as they sing from prominent places to attract a female.

PARAKEET, KINGFISHER AND DIPPER

Another motley collection of birds: two linked by their gaudy colours, and the other a passerine that thinks it is a waterbird.

Parakeets really have no place in a book about British birds; apart from the fact that they have somehow become an established part of our avifauna during the past six decades or so, descended from escapees in south-eastern England. The original birds have thrived, bred and spread, and the current population is more than 17,000, showing how well the species has adapted to its new surroundings. Originally from northern India, the **Rose-ringed** (or **Ring-necked**) **Parakeet** is a highly adaptable species, able to withstand extreme cold and to exploit artificial and natural food resources, including food put out for the birds by us. Its favourite habitat is large wooded parks, where it nests in holes in trees (possibly threatening native hole-nesting species such as the **Jackdaw**, **Stock Dove** and **Starling**). Parakeets are easy to see, thanks to their habit of flying in flocks while uttering noisy high-pitched contact calls. If you spend time watching them you'll see they are fascinating birds, exhibiting the agility, intelligence and sophisticated social life that make parrots such an interesting family. At dusk they fly overhead in large flocks, on their way to a communal roost.

The **Kingfisher** is even more striking than the parakeets; indeed it has no rival for the position of our most colourful native species. A much smaller bird than many people expect, it can also be elusive, often only seen as it flies away in a flash of blue and orange. However, find a regular site and you may be lucky enough to get excellent views of the bird

▶ *A familiar sight on bird-feeders in the London suburbs, the Rose-ringed Parakeet is exceptionally agile and acrobatic*

feeding by plunging into the water for small fish and other aquatic life. Kingfishers nest in holes in sandy banks, and although the nest itself is hidden underground the birds can be observed going to and fro.

The **Dipper** is unique: a songbird that hunts for its food underwater. Superficially resembling a huge wren, its aquatic habits and black-and-white coloration have earned it the folk-name of 'water ouzel'. Dippers prefer fast-flowing streams and rivers, and once you have found them will provide hours of entertainment as they fly to and fro, perch on rocks bobbing up and down, or plunge beneath the water in order to catch their aquatic food. They nest underneath the banks in a crevice or hole, and the young leave the nest before they are fully fledged, and are fed by the parents.

▲ *The Kingfisher is one of the most beautiful of all Britain's birds, though it is often hard to see as it can be quite shy*

▼ *The Dipper is the only British songbird that can dive under water to feed*

WOODPECKERS

Britain has only three native species of woodpecker, compared with 10 in continental Europe – and until recently (when Great Spotted colonised for the first time) none in Ireland! The reason is simply that as poor flyers and largely sedentary species, woodpeckers only spread north and westwards slowly after the end of the last Ice Age. Three pioneering species (plus the Wryneck, now almost extinct in Britain) managed to cross the land bridge to England before the sea cut us off from the continent, while none quite managed to reach Ireland before it too became an island. Woodpeckers excavate a new nest hole every year, so the old, disused holes are vital to other species of hole-nesting bird including Starling, Jackdaw and the introduced Rose-ringed Parakeet.

Two out of three British woodpeckers are relatively common and easy to see. Our largest species, the **Green Woodpecker**, is a bird of relatively open grassy areas with scattered trees, such as large parks, where it can be seen on the ground, feeding on its favourite food of ants. It also visits large open lawns, but is quite shy and will always be on the lookout, so don't approach too closely. This species drums less often than its 'spotted' relatives, and is best located by its far-carrying laughing call, which earned it the folk-name of 'Yaffle', and is supposed to forecast rain.

Of the two black-and-white woodpeckers, the **Great Spotted** is by far the commonest and most frequently seen. It drums more often than the other two species, and can also be detected by its penetrating 'chip' call. Great Spotted prefers more dense woodland than the Green, but is also a frequent visitor to gardens, where it will readily feed on bird feeders. It can also prey on birds, raiding nests and nestboxes for chicks. Look out for its undulating flight pattern.

▼ Our largest member of its family, the Green Woodpecker often feeds on the ground on ants

▲ *The Lesser Spotted Woodpecker is our smallest, rarest and most elusive woodpecker*

The smallest European woodpecker, **Lesser Spotted** is far more elusive. Indeed, it behaves much more like a passerine than a woodpecker, creeping around the topmost branches of a tree like a Treecreeper or Nuthatch (which it resembles in size if not appearance). Your best chance of seeing it is in early spring when birds are calling and drumming; or in winter, when they often tag along on the edge of a tit flock as it passes through woodland in search of food. Nevertheless you rarely get good views of this species.

◀ *Great Spotted Woodpeckers are now a familiar sight in parks and gardens, including those in towns and cities*

▲ *The Swift is the most aerial of all British birds, spending virtually the whole of its life in the air*

▼ *House Martins build their nests out of tiny pellets of mud*

SWIFT, SWALLOW AND MARTINS

Despite a superficial similarity, Swifts are not at all closely related to hirundines, the group that includes Swallow and martins. Nevertheless I have treated them together for the sake of convenience, as they share similar behavioural traits.

Swifts are among the most incredible birds in the world. They are the ultimate flying machines, able to stay aloft for months (even a year or more), and only landing to breed. Even then a Swift will never intentionally land on the ground, as its tiny legs placed at the rear of its body mean that it can never take flight again under its own steam. As a result Swifts nest in crevices in high buildings, where they can cling easily to the walls and 'fall' back into the air. They return en masse in late April or early May, appearing in huge numbers over reservoirs and marshes where they can find plenty of insect food to refuel after their long journey from tropical Africa. Then they disperse to towns and cities throughout the country, where they spend the evenings chasing each other across the skyline, uttering the screams that gave them the folk-name 'the devil bird'.

Swifts usually begin nesting in May, but if bad weather arrives will often disappear for days or even a week or more, their young staying torpid in the nest until the adults return and begin feeding them again. Once the young leave the nest they are entirely independent of their parents, foraging for tiny flying insects on their own. The sight of Swifts flying high for insects on a warm summer's evening is one of the most characteristic scenes of summer life, and when they suddenly disappear in August the urban landscape seems a poorer place without them.

▲ *Sand Martin (left) and Swallow (right) are both summer visitors to Britain, spending the winter in sub-Saharan Africa*

Of the three true hirundines, the **House Martin** is the Swift's urban and suburban companion, nesting under the eaves of houses. As its name suggests, it has adapted brilliantly to live alongside humans, having originally nested in caves and on the sides of cliffs. The birds arrive back in late April, check out their nesting sites, then often disappear for a week or so to feed on nearby lakes or reservoirs. Once they return they get down to repairing and rebuilding their old nests, or starting new ones from scratch, using mud collected from a nearby stream, building site or farmyard. They will also readily take to artificial nestboxes, especially useful when a supply of mud is not easily available. They are wonderful birds to observe as they go to and fro to feed their hungry young, though their noisy calls can wake you up early. On fine summer evenings (and indeed well into the autumn) they can be seen hawking for insects, sometimes high in the sky.

Sand Martins are birds of gravel pits, quarries, undercliffs and riverbanks; anywhere where they can burrow into a soft bankside and raise their families. They are very early arrivers, often here by mid-March, and like other hirundines they feed over water where there is a plentiful supply of small flying insects. If you find a colony it will give you hours of fascinating observation.

The **Swallow** is not only a sign of summer, but also one of our best-loved birds. Its graceful flight and habit of nesting close to human habitation (especially in farmyards) makes it a familiar and welcome summer visitor. Swallows will build their nests in buildings and barns, and can easily be detected by their noisy calls. They tend to hunt lower than martins, often skimming the grass in between farm animals that attract insects. In late summer and early autumn look out for gatherings of Swallows and martins as they perch on telegraph wires prior to migration, filling the air with their contact calls.

LARKS, PIPITS AND WAGTAILS

Three groups of birds, two of which (larks and pipits) are not related but superficially resemble each other; and two of which (pipits and wagtails) are related but at first sight may appear quite different. However, a closer look reveals similarities in structure and behaviour.

There are three species of **larks** found in Britain, two breeders and one wintering. Of these by far the most common is the **Skylark**, one of our best-known and best-loved birds, which sadly in recent years has undergone a decline in numbers. Nevertheless it is still widely distributed in the countryside, in various habitats from lowland farms to upland moors. It is most famous for its extraordinary song-flight, during which the bird rises so high in the sky it may become almost invisible, pouring its heart out in a continuous song before plummeting down to earth. Once on the ground, it generally lands some distance from its nest and runs the last few metres, making it hard to discover the location of the nest. Outside the breeding season Skylarks form large flocks which range over stubble fields to feed. They may also undertake local movements, especially during hard weather in winter.

The **Woodlark**, despite its name, is actually a bird of open heath with a few scattered trees. Once threatened with extinction as a British breeding bird, it has made a remarkable recovery in recent years, thanks partly to the felling of conifer populations and the creation of more open heath. It has a beautiful song, often uttered early in the spring from a song-post on a bush or tree.

▼ *The Skylark (left) and Woodlark (right) both have beautiful songs that are widely celebrated in poetry and music*

The Meadow Pipit is one of the most widespread of all Britain's breeding birds, found in a wide range of upland and lowland habitats

The **Shore Lark** is a winter visitor to Britain, and as its name suggests is found on sandy shores and coastal saltmarshes. It usually forms loose flocks, often associating with Skylarks and **Snow** and **Lapland Buntings**.

Our three breeding species of **pipits** exploit a wide range of habitats. The **Meadow Pipit** is highly adaptable and catholic in its choice of habitat, and is a typical bird of moors, heaths and grassy fields. In winter it also comes down to the coast, where it may be confused with its much more specialised relative the **Rock Pipit**. This species is unique amongst British birds in that it is the only songbird to have an exclusively coastal breeding distribution. **Tree Pipit** is more a bird of heathland and the edge of woodland plantations, performing its song-flight from a high perch, and launching itself into the air before parachuting down again.

Finally, the **Water Pipit** is also unique: it is the only British songbird to arrive from the south as a non-breeding visitor. Breeding in the high mountains of Europe, it spends the winter in a range of habitats including riverside, watercress beds and marshes. All pipits are gregarious birds, often forming loose feeding flocks.

Of our three **wagtail** species, the **Pied Wagtail** is by far the most widespread and adaptable. It seems to love concrete and tarmac, often being the only bird to walk around this unpromising 'habitat', apparently picking up tiny morsels of insect food while bobbing its tail. It will also regularly visit garden lawns. Pied Wagtails roost in some unusual places, such as trees in the centre of city squares and shopping centres, or factories, where they take advantage of extra warmth provided by industry or retail outlets. In flight they give a characteristic 'chis-ick' call.

▲ *Pied Wagtails are one of the most familiar of all town and city birds, found in parks, gardens and roadsides*

▼ *Despite its bright lemon-yellow colour, this is actually a Grey Wagtail*

Yellow and **Grey Wagtails** can be confused on first view, as both have plenty of yellow in their plumage; however, they are structurally quite different with the Grey much longer-tailed. Yellow Wagtails are summer visitors, generally seen on marshes and flooded fields (an increasingly scarce habitat), while Grey Wagtails are resident, and are closely associated with running water such as streams or rivers, though they can also be found round the edge of reservoirs and ponds. Both feed in the characteristic manner of their family.

THRUSHES AND CHATS

In this group I include a dozen familiar and not-so-familiar species, including six 'true' thrushes, and six smaller species known as 'chats', which are closely related to each other. Chats have much in common behaviourally with the thrush family, although recent DNA studies show they are actually more closely related to the flycatchers. A seventh chat, the Robin, is dealt with in the following section.

Of the thrushes, the most familiar must surely be the **Blackbird**. Found throughout Britain apart from some upland areas (where it is replaced by the **Ring Ouzel**), it is a common and familiar resident in towns, suburbs and the countryside, and is particularly partial to nesting in gardens. Early in the year listen out for the deep, fluty song of the male, or the angry chattering call as it chases away potential rivals. Blackbirds are highly territorial, and the male will continue singing even when he is feeding chicks. The female is a much less obvious bird, which generally feeds by creeping around in the undergrowth. In autumn Blackbirds often feed on windfall apples, alongside other thrushes.

The **Ring Ouzel** is the upland equivalent of the Blackbird, so is mainly seen in northern and western Britain. It is a summer visitor, and may sometimes be found on migration in lowland habitats, which it tends to visit year after year. Like the Blackbird it is an excellent songster, perching high on bushes and rocks in order to deliver its song.

The **Mistle Thrush** and **Song Thrush** are often confused, though their size and plumage details are fairly distinctive. Behaviourally, too, they differ: the Song Thrush prefers more wooded habitats and

▼ *The Ring Ouzel is the highland version of its cousin the Blackbird, found on moors and mountains in northern and western Britain*

gardens, generally singing from the top of roofs; while the Mistle Thrush is a bird of open parkland with scattered trees which it uses as song-posts. The Mistle Thrush also has the reputation of singing before and during bad weather, which earned it the country name of 'Stormcock'. Outside the breeding season Mistle Thrushes often gather in flocks, calling as they fly overhead in search of open areas of grass to feed. In winter they defend berry bushes against all-comers. Song Thrushes tend to be shyer and more solitary, leaving gardens for nearby wooded areas. Song Thrushes are avid consumers of snails, breaking their shells on a large stone known as an 'anvil' in order to get at the juicy contents.

In autumn these native species are joined by our two 'winter thrushes', the **Redwing** and **Fieldfare**. In some ways these are the northern equivalents of the Song Thrush and Mistle Thrush respectively. Both species travel in loose flocks, sometimes with each other, and raid berry bushes or feed in the open in fields. They also migrate in flocks, calling to each other as they pass overhead, often at night.

Our first two species of chats each exploit slightly different habitats. **Whinchats** are birds of upland areas such as moors, though on migration they can be seen almost anywhere. **Stonechats** are more associated with gorse and heathland habitat, but those birds that do not migrate will spend the winter in more general habitats such as near reedbeds. Whinchats all migrate to sub-Saharan Africa for the winter.

▶ *The Whinchat was once found throughout much of Britain, but has now been driven back into the uplands due to habitat loss and lack of food*

▲ *The Redstart (top) and Black Redstart (bottom) both have red tails – the word 'start' derives from an Anglo-Saxon word meaning 'tail'*

A close relative, the **Wheatear**, is also a migrant, arriving back as early as March. It is also a bird of moorland areas, though is often found on beaches and other coastal areas during migration, feeding on the ground. Its name has nothing to do with wheat – it is a corruption of an Anglo-Saxon word meaning 'white-arse', which when you see the bird flicking its tail and wings to reveal its white rump seems very appropriate.

The names of two other chats, the redstarts, also derive from Anglo-Saxon, 'start' meaning 'tail'. The reddish-brown tail is an obvious identification feature. The **Redstart** is a summer visitor, and mainly found in mature broad-leaved woodland, where its song is often the

▶ *The Nightingale is justifiably the most famous songster in Britain, though its population has declined in recent decades*

first clue to its presence. It can be a shy bird, but may be seen visiting its nest hole. Like other chats and thrushes it may be found in a wider range of habitats on migration. Its scarcer cousin, the **Black Redstart**, has more peculiar tastes in terms of habitat. On the continent it is a bird of rocky slopes, cliffs and towns, but in Britain it prefers to breed in industrial areas such as building sites and even nuclear power stations, having colonised via bombsites after the Second World War. In autumn and winter birds often disperse to coastal areas, and are generally found near water where they can find insect food.

Another chat is the bird celebrated more than any other by poets and writers: the **Nightingale**. It is a paradoxical bird: one with a stunning song yet a drab plumage and shy, retiring habits. Your best chance of seeing one is when the males arrive back in late April and early May and begin singing to defend a territory and attract a mate, often sitting right out in the open. Once the females arrive and they pair up the males become incredibly shy, singing from the centre of dense foliage in their woodland habitat or on the edge of heaths. At this stage they may simply be impossible to see, but they sound wonderful. As you might expect they do sing at night, though they often give plenty of song during the daylight hours as well.

ROBIN, DUNNOCK AND WREN

These three familiar garden species are among our commonest and best-known birds, and each provides a perfect opportunity to study fascinating behaviour at close quarters, often from the comfort of your home.

The **Robin**, which is a species of chat, regularly wins polls of Britain's best-loved bird, yet is perhaps the most aggressive of all our songbirds, with rivals fighting viciously, sometimes even to the death. Once they have established their territory they will defend it violently, with both sexes being strongly territorial. Robins nest in a wide variety of places, including bizarre locations such as toilet cisterns, under the bonnet of vehicles, and at the top of drainpipes. They will also readily take to nestboxes. Once hatched the young may appear in your garden, looking quite unlike their smart red-breasted parents. In autumn our native Robins are joined by continental immigrants, which often appear in quite large numbers on the east coast during 'falls' brought about by the right weather conditions. Robins are one of the very few birds that sing all year round (even at night, which often leads to out-of-season claims of Nightingales). They do so in order to defend a territory in autumn and winter; and unusually amongst British birds, female Robins also hold a territory and sing at this time of year.

The shy, unobtrusive **Dunnock** has a torrid sex-life. Both sexes are often polygamous, having a main 'alpha' partner but also a regular

▼ *The Robin has been closely associated with Christmas and the Festive Season for more than a century, partly because of its confiding habits*

▶ *The Dunnock was once known as the 'Hedge Sparrow', even though it is actually a member of the accentor family*

'beta' partner on the sly, and a male may follow his mate about to try to prevent her from mating with other males. He will even use his bill to remove sperm from her cloaca. Singing male Dunnocks often pick elevated and prominent song-perches. Otherwise they tend to creep about flower borders like little mice, occasionally venturing out onto the lawn to pick up seeds dropped from a feeder or bird table.

Wrens are also very territorial, and like the Dunnock the male becomes highly visible in early spring, often singing right out in the open, with an incredibly loud song for such a small bird. Male Wrens often build several proto-nests, from which the female chooses the best and completes it before laying her eggs. Outside the breeding season Wrens hop about unobtrusively in search of insect food, and in harsh winter weather will often roost in empty nestboxes, with several birds huddling together for warmth.

▶ *The Wren is now Britain's commonest bird by far, with at least seven million breeding pairs*

WARBLERS

There are 13 regular breeding species of warbler in Britain, and two species of warbler-like 'crests'. Most are summer visitors, and all 15 can be grouped into three categories by habitat.

Wetland warblers

This category includes five species that are habitually associated with wetland areas during the breeding season, though the actual habitat varies from species to species. The classic 'reedbed' species is the **Reed Warbler**, which as its name suggests lives almost exclusively in reeds, where it makes its nest by weaving grasses around the reed stems. It can be heard delivering its distinctive, repetitive song from mid-April, though birds may be reluctant to show themselves, especially in windy weather when they tend to stay low down in the reeds.

Its close relative the **Sedge Warbler** shares its reedbed habitat, though tends to sit on a more prominent perch such as a small bush in order to deliver its song. Males also launch themselves into the air in a song-flight, parachuting down to their perch as they sing.

Very similar in appearance to the Reed Warbler, the **Marsh Warbler** is much rarer in Britain, and prefers damp wooded habitat on the edge of ponds and streams. It is a highly accomplished mimic, and the song may include an incredible variety of snatches from the songs of other birds – not just British species but those encountered on its African winter quarters as well.

Grasshopper Warbler is so-named after its extraordinary reeling song, which sounds like a cross between an insect and a fishing reel. It too prefers a less 'wet' habitat, and is often found in quite dry bushy

◀ *Reed Warblers come north from Africa each spring to breed in Britain's reedbeds*

▲ *Although Blackcap (top) and Garden Warbler (bottom) are different in appearance, they have remarkably similar songs*

areas near the edge of marshy ones. It tends to sing most at dawn, dusk and even through the night, and although not usually easily visible, once discovered may allow quite a close approach.

The final 'wetland' species, **Cetti's Warbler**, is a fairly recent colonist to Britain, and unlike most other warblers it is a resident species. Its presence is usually noted when it sings its incredible explosive, richly melodious song. Occasionally the bird may show itself, sometimes even giving good views, but it is usually frustratingly hard to locate.

Scrub warblers

This category covers warblers of the genus *Sylvia*, a largely Mediterranean group of birds of which five species breed in Britain.

The most familiar of these is probably the **Blackcap**. Not only is it a common summer visitor, found in a variety of rural and suburban habitats including large gardens, but in recent years a population from central Europe has also begun to spend the winter in Britain, often visiting gardens in search of food. The Blackcap is a fine songster, often compared with the Nightingale, though without the same range and beauty. Garden visitors have become quite adaptable, feeding on bird tables and feeders, though the birds that are here in summer mainly live in woodland areas.

Its sibling species, the **Garden Warbler**, is not a very frequent garden visitor, despite its name. It prefers fairly open woods, but because of its retiring habits, unmarked plumage and very similar song to the Blackcap it is often overlooked. The song tends to be faster, less varied and lacking the fluty tones of its commoner relative.

The two whitethroats also form a species pair, separated by their different choice of habitat. The **Common Whitethroat** is found in a range of habitats including heath, hedgerows, farmland and parks, and draws attention to its presence by singing its rapid, scratchy song either from a prominent perch or from mid-air in a steeply rising song-flight. After the breeding season it may be found feeding on berries in preparation for its long journey to sub-Saharan Africa.

The **Lesser Whitethroat** is far more elusive than its cousin, and its presence may only be detected by its sharp, dry call or fairly distinctive rattling song, emanating from dense scrub or bushes.

▼ *The Lesser Whitethroat is one of the most skulking British warblers, though males do sometimes sing out in the open*

▲ *Chiffchaff (left) and Willow Warbler (right) are best told apart by their very distinctive songs*

The final member of this group, the **Dartford Warbler**, is a truly resident species, on the northern edge of its range in Britain. It prefers gorse and heathland as a breeding species, often singing prominently from the top of bushes, especially during fine spring weather. Outside the breeding season some birds disperse to less specialised habitat, including bracken and parkland.

Leaf warblers and crests

The three British 'leaf' warblers are well known to be so similar that they were not told apart until the 18th century, when Gilbert White distinguished between the species. In fact with modern identification techniques and optics (which of course White did not have) the three species are fairly straightforward to identify. They also display quite different behaviour.

The **Willow Warbler** and **Chiffchaff** are the two most similar species of the three. As well as their very distinctive songs, they also exploit rather different habitats, with Willow Warbler preferring heathland as well as mixed woodland, and Chiffchaff often found on the edge of woodland. Both are active feeders, and on migration can be found in unusual habitats such as coastal scrub and even gardens. Chiffchaffs also winter in good numbers, especially in the milder south-west, and are often found near water where they can be seen hunting for insects.

The **Wood Warbler** is a larger, brighter bird and has a very distinctive singing behaviour: shivering its wings in time with its delightful song, then flying a short distance to another part of its territory and starting to sing again elsewhere in the woodland canopy.

▲ *The Wood Warbler is a summer visitor to western Britain, often found in oakwoods and other ancient woodlands*

The two 'crests' are very active, tiny birds (the smallest in Europe), constantly on the move in search of tiny insects. Both are found in a variety of woodlands, though **Goldcrests** have a preference for conifers and will often hunt deep inside the foliage and so can be hard to see. Listen out for their distinctive calls and song, which are often the best way to find them. **Firecrests** tend to move on more quickly from tree to tree than Goldcrests, and are often found in autumn and winter near the coast, or near water, where a milder climate encourages more insects. Both species will follow flocks of tits in winter.

◀ *The Goldcrest is Britain's smallest bird, weighing barely five grams – the same as a sheet of A4 paper or a 20 pence coin*

FLYCATCHERS

Two species of flycatchers, both summer visitors, breed in Britain.

The **Spotted Flycatcher** is the most widespread, found in a range of lowland habitats including rural gardens and woods. Like all members of its family it lives up to its name, sallying forth from a branch or twig to catch small flying insects in its bill, before returning to its perch. Spotted Flycatchers are among the latest migrants to return, and once here they nest in crevices in walls, open ledges or in tree-forks.

The **Pied Flycatcher** is, by contrast, a hole-nester, and prefers mature mixed woodlands, mainly in the western half of Britain. Like its relative it also flycatches for food. Both species may turn up in unusual places during migration.

▶ *Pied Flycatchers readily take to breeding in nestboxes, especially where holes in trees are in short supply*

TITS, NUTHATCH AND TREECREEPER

This set of woodland species often spend time in close proximity, especially during autumn and winter when they will form mixed flocks comprising several different species, to hunt for insect food. During this time of year the wood may seem empty until you hear some tiny, high-pitched contact calls by which these little birds stay in touch with each other and signal a new food resource to their fellow travellers.

There are six 'true' **tits** in Britain, together with two similar species that also bear the name. The true tits include some of our commonest garden birds together with much scarcer species. **Great**, **Blue** and **Coal Tits** are all common and widespread, and often visit gardens to supplement natural food sources, or to nest, often in artificial nestboxes. Their feeding behaviour delights many a home-owner as they squabble with each other to get the best position on the hanging feeder, then extract a morsel of peanut or energy-rich seed. Great Tits are the top dogs in the hierarchy, though Blue Tits often sneak in cheekily under their bills, as it were. Coal Tits tend to hang back and are a bit shyer.

Outside the garden all three species are found in mixed woodland, with Coal Tits also having a liking for coniferous forest. In the woods they are joined by **Marsh** and **Willow Tits**, a sibling pair of species that look very alike. Marsh Tits also visit gardens from time to time, but mainly inhabit dry wooded areas, while Willow Tits prefer damp

▼ *The Blue Tit is one of the commonest and most familiar British breeding birds, especially in gardens*

woods, often close to water. All these species may join tit flocks in winter, though Willow do so more rarely; and all nest in holes or cavities in trees. The final 'true' member of the family, **Crested Tit**, is confined to the Scottish pine forests, though there it will behave in true tit fashion, nesting in holes in trees and joining feeding flocks, even coming to artificial feeders where provided. Willow and Crested Tits both excavate their own nest holes in rotten wood; unusual behaviour for such tiny songbirds.

Another species, **Long-tailed Tit**, comes from a different family, but to all intents and purposes behaves like the other tits, especially when feeding. Long-tailed Tits often travel in flocks of up to a dozen or more birds, usually related to each other, calling constantly as they move acrobatically through the foliage. If you are patient and still they will

▼ *Although superficially similar, given good views Coal (top) and Marsh (bottom) Tits are easy to tell apart*

▲ *Crested Tits (left) are confined to the forests in and around Speyside, while the Treecreeper (right) can be found in woodlands throughout much of Britain*

often come very close. Unlike other tits they make their own nest out of moss and lichens, which looks like a ball with a small entrance hole. All these woodland tits will often respond if you try 'pishing' – making a hissing or squeaking noise – which seems to arouse their curiosity and brings them closer.

An eighth British 'tit' is not a tit at all, but in its own unique family. The **Bearded Tit** (also known as the **Bearded Reedling**) is a bird of reedbeds, where it spends virtually its whole life. It is a beautiful and elegant bird, often seen in small parties which are usually found by listening for their distinctive, dry metallic 'pinging' calls. On a calm day you may get good views as birds climb to the tops of reeds; but if it is windy you don't stand much of a chance.

The **Nuthatch** and its relatives are unique: the only birds that can walk down, as well as up, a tree trunk. They do so by using their formidable claws. The Nuthatch is rather like a miniature woodpecker in appearance and habits, moving up and down tree trunks to find food, and nesting in holes, which it adjusts to fit if they are too large by plastering mud around the entrance. Nuthatches live up to their name (it means 'nut hack') by opening the shells of nuts by wedging them into a crevice in a tree and then hammering them with that powerful bill.

The **Treecreeper** is often seen alongside Nuthatches and tits. It is an easily overlooked bird, creeping like a mouse around the branches and twigs to find insect food, and only reluctantly flying to the next tree. It habitually climbs in spirals, going around the back of a trunk or branch, and then reappearing around the front again.

SHRIKES

Two species of shrikes are occasionally seen in Britain: one irregular breeder now usually seen as a migrant; and another regular but very scarce winter visitor.

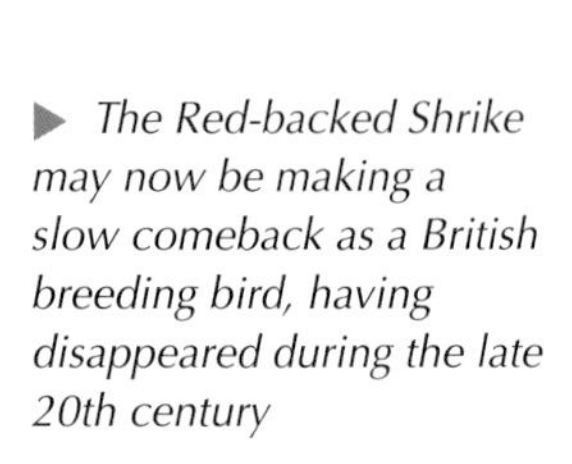

▶ *The Red-backed Shrike may now be making a slow comeback as a British breeding bird, having disappeared during the late 20th century*

The **Red-backed Shrike** earned the country name of 'butcher bird' from its unpleasant but fascinating habit of impaling its prey on thorns. Sadly it was lost as a British breeding species during the 1980s, probably due to a combination of modern farming methods reducing the number of large insects, and climate change, which brought wetter summers at the time of the species' decline. However, in recent years a handful of pairs have bred in southern England, and with global warming potentially leading to a more benevolent summer climate for the species, it could return permanently. Its larger relative, the **Great Grey Shrike**, may be seen in winter at regular haunts on heaths in southern and eastern Britain, though has declined in recent years. It too sits on high, prominent posts and hunts for food by diving down on small mammals and birds.

▼ *The Great Grey Shrike is a scarce and much sought-after winter visitor to Britain*

STARLING

One of the commonest and most taken-for-granted of British birds, the Starling is also one of our most fascinating in terms of its habits.

◀ *Starlings are often taken for granted, even though they are both attractive and fascinating*

Starlings are intelligent and sociable birds, with complex breeding and flocking behaviour. In the breeding season males sing from perches on trees and especially roofs, delivering a song extraordinary for its mimicry – not just of other birds but also of car alarms, mobile phones and a host of other mechanical objects.

Outside the breeding season roving flocks of Starlings feed in gardens (especially on open lawns and bird tables), playing fields and farmland, probing short grass for invertebrates. They also come together – especially in winter – at dusk, forming huge, noisy flocks. These were once a common sight in many urban areas as well as the open countryside, but in recent years the species has declined severely, and not as many sites host large flocks as they once did.

If you do know a site, plan to visit an hour or so before dusk, and marvel at the birds' behaviour as they swirl around in the sky, with small groups joining every minute or so, before they finally settle on a building or in the trees. A truly amazing experience.

CROWS

The eight British breeding species of crows are among the most intelligent of all birds, and display a wide range of fascinating behaviour. Crows are especially curious birds, known for their ability to learn innovative new behaviour; and also, less endearingly, for preying on other birds, their chicks and eggs.

There are three species of large, black crow. The **Carrion Crow** is the classic member of the group, with its adaptability and omnivorous habits making it widespread throughout Britain, apart from the extreme north and west where it is replaced by a very similar but black-and-grey species, the **Hooded Crow**. Both are the duckers and divers of the bird world, always on the lookout for an opportunity, though this often involves muscling in on smaller birds. Both often gather in large flocks to feed or roost, and will also mob birds of prey without compunction.

The **Rook** is a gentler bird, usually found in more rural areas where it gathers to breed early in the year at its famous rookeries. Rooks are also famous for their 'tumbling' behaviour during windy days in autumn, which is supposed to foretell unsettled weather to come. The biggest crow of all is the **Raven**, a magnificent bird with glossy black plumage and a commanding stature. Ravens are birds of the uplands, and will nest early in the year on prominent crags, from where they survey their little world. However, in recent years they have also made a comeback in lowland Britain, thanks to a reduction in their persecution. They are also supreme flyers, often courting high in the air.

▼ *The all-black Carrion Crow is the commonest and most familiar member of its family in much of Britain*

▲ *The Raven is the largest British crow, and has recently made a comeback, spreading back to south and east Britain*

Two smaller black crows, the **Jackdaw** and the **Chough**, are also fascinating birds. Jackdaws are real comedians, often associating with their larger relatives in search of food on farmland or in open parks. Their characteristic call is what gave them their name, as it is for their rarer relative the Chough (originally pronounced 'chow', which mimics their call). Choughs are also comical birds to watch as they walk around on short grass on their huge red feet, poking that long, red, decurved bill into the soil to get at their food.

◀ *The Jackdaw is the smallest 'all-black' crow, told apart by its small size, grey nape and beady pale eye*

▲ *The Jay may be bright and colourful, but it is also very shy, so can be overlooked*

The remaining two members of the crow family are more colourful, or at least strikingly patterned, than their mainly dark relatives. The **Jay** is a familiar, if rather shy, garden bird, which often hangs about in trees and bushes before swooping quickly down to take food, either from a bird table or a nest. In autumn numbers are boosted by continental immigrants.

The **Magpie** is often portrayed as a villain because of its habit of taking eggs and young chicks from songbirds' nests, but like all predators its numbers are governed by its prey, not the other way round. Therefore although its behaviour can seem cruel (though in reality it is only doing what Magpies have to do to feed their own young) it has no lasting effect on songbird populations. Magpies are sociable birds, usually travelling in pairs or small groups, which are supposed to bring the observer good luck (seeing a lone bird, conversely, is said to bring bad luck).

SPARROWS AND BUNTINGS

These two families of seed-eating birds are closely related, and share many aspects of their behaviour with each other, so I have treated them together for the purposes of this book.

▲ *The House Sparrow has suffered a major decline in the past fifty years, and has now vanished from many of our urban areas*

▼ *The Yellowhammer is a member of the bunting family, found mainly on farmland*

Our two species of **sparrows** have both declined strongly in recent years, due to modern farming methods and various other reasons. **House Sparrows** were once so common as to be virtually ignored, but since their decline birders have begun to study them in greater depth. One possibility is that being sociable birds they need several pairs in an area to stimulate each other to breed; and because of declining numbers there are simply not enough birds in some areas to sustain a viable breeding population.

If this is the case we may see the species suffer the same catastrophic decline as its more rural relative, the **Tree Sparrow**. The population of this species has plummeted by more than 90%, partly due to modern farming methods not leaving enough grain on the fields for the birds to eat in winter. Both species will join with other seed-eating birds outside the breeding season. Tree Sparrows mainly breed in holes in trees, while House Sparrows are more adaptable, nesting in holes in buildings as well as trees.

Five species of **buntings** regularly breed in Britain. Three are widespread, while the other two are confined to particular areas of the north and south. The **Yellowhammer** is the most common and widespread, though like all farmland species it has declined in recent years. It is fairly easily seen where present, perching on hedgerows or bushes and singing its famous 'little-bit-of-bread-and-no-cheese'

▲ *The Corn Bunting may look rather dull, but it has a very interesting sex-life, with males pairing with several females at once*

song. In winter it joins forces with other buntings and finches to feed on stubble fields. Its much rarer and more localised relative, the **Cirl Bunting**, is confined to south-west England where it requires a traditional farmland habitat all year round. Its behaviour is very similar to its commoner relative.

The other common species of bunting are both farmland birds to a degree. The **Corn Bunting** is a common sight in some areas of lowland Britain where it perches on wires or poles to sing its characteristic 'key-jangling' song. However, it has also declined and may be absent from former haunts. Its breeding behaviour is fascinating: males pair with several females, and have to work hard to keep away rivals. The **Reed Bunting**, as its name suggests, is more a bird of wetlands, though it also uses farmland to breed. Reed Buntings have begun to visit gardens in recent years, even feeding on bird tables.

The remaining British bunting is confined as a breeding bird to the high Arctic–alpine habitat of Scottish mountains, but in winter flocks gather around our coasts. The **Snow Bunting** is a real specialist, able to breed farther north than any other small bird; though in autumn it migrates south to milder climes. Winter flocks feed on shingle beaches and saltmarshes, appearing quite inconspicuous while feeding, but showing an explosion of white when they take off. Snow Buntings can be very approachable in winter, especially when feeding on leftover food near the restaurant on the high tops of the Cairngorm mountains in Scotland.

FINCHES

Finches form a large family of more than 100 species, yet just 13 species breed in Britain – and that includes the taxonomically dubious Scottish Crossbill, which may prove to be less than a full species, and the Parrot Crossbill, which is an irruptive breeder, far commoner in some years than others.

The remaining 11 include some of our commonest and best-known birds, of which the **Chaffinch** is the most widespread and numerous. Like other finches it is a seed-eating specialist, able to eat a wide range of seeds, though like many seed-eaters it feeds its young on soft-bodied insects. Chaffinches form large flocks outside the breeding season, often associating with other species including their northern equivalent the **Brambling**. A rare breeding species here, Bramblings are abundant in northern Europe, and variable numbers arrive each autumn to seek out their favourite food of beechmast.

The other common finches include a generalist, the **Greenfinch**, and a specialist, the **Goldfinch**. Greenfinches eat a wide variety of seeds, and are particularly partial to artificial feeders containing either sunflower seeds or peanuts. As a result they are frequent garden visitors, and often stay to breed, the male performing his attractive display flight while singing as he flies. Goldfinches have a needle-sharp bill, ideal for prising the tiny seeds out of plants such as teasels, which they love. Listen out for the birds' tinkling calls as they fly overhead.

A much more recent garden visitor is the **Siskin**. Once confined to coniferous forests, mainly in northern Britain, the Siskin spread southwards a few decades ago and quickly adapted to feeding in gardens, where it is now a regular visitor. Siskins are lively little birds, which outside gardens tend to flock together near water, where they often join **Lesser Redpolls** feeding on alder cones. Lesser Redpolls will also visit gardens, and otherwise are generally found in damp woods.

▼ The Goldfinch is one of our most colourful and attractive songbirds, with a delightful, tinkling song

▲ *This male Linnet is sporting his colourful breeding plumage, with pink patches on the forehead and breast*

In the breeding season Siskins head north to breed in the pine forests of Norway and Scotland, while Lesser Redpolls breed on heathland.

Another 'pair' is the **Linnet** and the **Twite**, the latter sometimes known as the 'Mountain Linnet' because of its preferred habitat. Linnets are birds of farmland and heath, flocking in autumn and winter and pairing up in spring when the male adopts his splendid breeding garb. Twite also form flocks in winter, heading away from their hilly breeding areas towards the coast, where they often join with Snow Buntings.

The three finches with the biggest bills and toughest appearance are the **Bullfinch**, **Hawfinch** and **Common Crossbill**. The Bullfinch uses its powerful bill to feed on fruit buds, though it also eats other berries and seeds. It generally appears in pairs or small family parties, and is a shy bird, often overlooked. But for shyness, the Hawfinch takes the prize: despite being our largest finch, with a bill so powerful it can crack cherry stones, it is hardly ever seen even when present in wooded habitat. Hawfinches are best looked for at known sites in winter, when they gather in small flocks and are easier to see, as the trees are bare. The Common Crossbill (and its relatives) has a unique bill, with the tips of the mandibles crossed over to enable it to extract seeds from pine cones and other coniferous fruits. Crossbills are nomadic, and are the only species which migrates just once a year, moving in late summer to new feeding areas in different parts of Europe – so one year there may be hundreds of birds, the next year none.

▶ *The Hawfinch is Britain's largest member of its family, but is sadly now in decline*

▼ *With his black, grey and bright pink plumage, the male Bullfinch is handsome and distinctive*

GLOSSARY

Aberrant Individual bird that shows an unusual or abnormal characteristic compared to others of its species: for example, a leucistic (unusually pale) or melanistic (unusually dark) specimen.

Adaptation Evolutionary development of a particular characteristic to fit a change in the environment, such as a change in migratory habits, or tendency to breed earlier in response to climate change.

Alien A species that has been introduced into an environment from abroad: for example, Canada Goose and Ruddy Duck, which originate from North America, or Rose-ringed Parakeet, which comes from Asia. Often also known as 'introduced'.

Altricial A bird that hatches in the nest and remains dependent on its parents for food and care (also known as nidicolous). This applies to all passerine birds, for example.

Circadian Relating to the rhythms of the day: for example, the patterns of being awake or asleep.

Crepuscular A species which usually is most active at dawn and dusk, such as Spotted Crake.

Diurnal Generally active during the day, as opposed to at night (nocturnal).

Eclipse A camouflaged plumage acquired by male ducks during the mid to late summer moult, when the birds often become flightless for a short time.

Feral A species or population that was originally kept by humans, but following accidental release is now living freely in the wild. Examples include Greylag Goose and Rose-ringed Parakeet.

Fledging The point at which (mainly) songbirds leave the nest and are ready to fly. Also used to describe the point at which precocial birds such as gamebirds or waders moult into their first proper plumage.

Habitat A particular place or area where creatures such as birds live: e.g. woodland, heathland. Characterised by distinctive vegetation and soil, and often also by climate.

Incubation The period during which a bird sits on its eggs, in order to provide the warmth necessary for development of the embryo.

Irruptive A species that periodically invades an area where it is not normally found, usually in order to find food, such as Crossbill, Waxwing or Nutcracker.

Jizz A birder's term for the 'general impression' given by a bird, even if it is too distant or badly lit for details of plumage and coloration to be detected.

Longevity The length of time that a bird lives.

Migratory A species or population which undertakes (usually twice-yearly) medium or long-distance movements between separate breeding and wintering grounds.

Morphology The external shape and form of a bird or other living organism.

Moult The process of shedding old, worn feathers and regrowing new ones.

Passerine The taxonomic order containing the perching birds; also known as songbirds.

Polygamous A mating system in which either one male breeds with several females (polygyny); or one female breeds with several males (polyandry).

Precocial A bird that can see, walk, feed itself, and in some cases swim soon after hatching (also known as nidifugous). Examples include ducks, gamebirds and waders.

Predator Any bird that hunts and kills other vertebrate animals in order to feed.

Race (See Subspecies)

Reintroduced A formerly native bird which, having disappeared from Britain (or part of Britain), has been brought back by human agency, such as the White-tailed Eagle in Scotland and the Red Kite in England and Scotland.

Sedentary A bird that spends all or most of its life in the same area; as opposed to migratory.

Subspecies A distinctive population of a particular species, which has diverged enough to be distinct from others of the same species, but is not yet so different as to be classified as a full species. Sometimes known as race.

USEFUL ADDRESSES

RSPB (Royal Society for the Protection of Birds)
The Lodge, Sandy, Beds SG19 2DL
Tel: 01767 680551
www.rspb.org.uk

The RSPB is Britain's leading bird and wildlife conservation organisation, with well over one million members. It runs more than 200 bird reserves up and down the country, and has a national network of members' groups. Members receive four copies of *Nature's Home* magazine each year, while new members receive a gift on joining. The junior arm of the RSPB, Wildlife Explorers, is for members aged 4–19.

BTO (British Trust for Ornithology)
The Nunnery, Thetford, Norfolk IP24 2PU
Tel: 01842 750050
info@bto.org
www.bto.org

The BTO offers birdwatchers the opportunity to learn more about birds by taking part in surveys such as the Garden BirdWatch or the Nest Record Scheme. BTO members also receive a bi-monthly magazine, *BTO News*.

WWT (Wildfowl and Wetlands Trust)
Slimbridge, Gloucestershire GL2 7BT
01453 891900
enquiries@wwt.org.uk
www.wwt.org.uk

The WWT is primarily dedicated to conservation of the world's wetlands and their birds. It runs nine centres in the UK, including the London Wetland Centre and the famous headquarters at Slimbridge in Gloucestershire. Members receive a quarterly magazine, *Waterlife*, and get free entry to WWT centres.

CJ Wildlife
The Rea, Upton Magna, Shrewsbury SY4 4UR
Tel: 0800 731 2820 (Freephone)
sales@birdfood.co.uk
www.birdfood.co.uk

CJ Wildlife (formerly CJ WildBird Foods) is one of Britain's leading suppliers of bird feeders and foodstuffs, via mail order. The company produces a free handbook of garden feeding, containing advice on feeding garden birds, and a catalogue of products.

WildSounds
Cross Street, Salthouse, Norfolk NR25 7XH
Tel: 01263 741100
www.wildsounds.com

Wildsounds is one of Britain's leading mail-order suppliers of bird books, CDs and other products.

Subbuteo Natural History Books Ltd
The Rea, Upton Magna, Shrewsbury SY4 4UR
Tel: 0870 010 9700
www.wildlifebooks.com
sales@subbooks.demon.co.uk

Subbuteo Books provides a fast, helpful and reliable mail-order service for books on birds and other aspects of natural history, including those on garden birds. Free catalogue available on request.

NHBS
1–6 The Stables, Ford Road, Totnes,
Devon TQ9 5LE
Tel: 01803 865913
www.nhbs.com

An online natural history, science and environment bookstore

FURTHER READING

The World of Birds
By Jonathan Elphick (2014, Natural History Museum)

Bird Sense
By Tim Birkhead (2013, Bloomsbury)

Collins Field Guide to Bird Songs and Calls
(book & 3 CDs)
By Geoff Sample (1996, Collins)

Bird Migration: a general survey
By Peter Berthold (2001, Oxford University Press)

The Migration Atlas
By Chris Wernham *et. al*
(2002, T & AD Poyser)

Bird Atlas 2007-11
By Dawn Balmer *et. al*
(2012, BTO)

INDEX

ACKNOWLEDGEMENTS

At Bloomsbury, many thanks to Jane Lawes and Nigel Redman for producing the revised edition of this book, and also to Susan McIntyre for designing the book, Krystyna Mayer for copyediting and Liz Shaw for proofreading.

As someone who has watched birds for virtually the whole of my life, I should like to thank all my companions in the field – both long-term friends and casual acquaintances – who have prompted my interest in bird behaviour over the years. These include Daniel Osorio, Neil McKillop, Bill Oddie, Nigel Bean, Nigel Redman, Jackie Follett, Rod Standing and Graham Coster.

And as always, to my wife Suzanne, whose ability to see what I often miss has opened my eyes to a whole new world of birds and their habits.